CLAY

The History and Evolution
of Humankind's Relationship
with Earth's Most
Primal Element

Suzanne Staubach

BERKLEY BOOKS
NEW YORK

The Berkley Publishing Group
Published by the Penguin Group
Penguin Group (USA) Inc.
375 Hudson Street, New York, New York 10014, USA
Penguin Group (Canada), 90 Eglinton Avenue East, Suite 700, Toronto, Ontario M4P 2Y3, Canada
(a division of Pearson Penguin Canada Inc.)
Penguin Books Ltd., 80 Strand, London WC2R 0RL, England
Penguin Group Ireland, 25 St. Stephen's Green, Dublin 2, Ireland (a division of Penguin Books Ltd.)
Penguin Group (Australia), 250 Camberwell Road, Camberwell, Victoria 3124, Australia
(a division of Pearson Australia Group Pty. Ltd.)
Penguin Books India Pvt. Ltd., 11 Community Centre, Panchsheel Park, New Delhi–110 017, India
Penguin Group (NZ), cnr Airborne and Rosedale Roads, Albany, Auckland 1310, New Zealand
(a division of Pearson New Zealand Ltd.)
Penguin Books (South Africa) (Pty.) Ltd., 24 Sturdee Avenue, Rosebank, Johannesburg 2196, South Africa

Penguin Books Ltd., Registered Offices: 80 Strand, London WC2R 0RL, England

While the author had made every effort to provide accurate telephone numbers and Internet addresses at the time of publication, neither the publisher nor the author assumes any resonsibility for errors, or for changes that occur after publication. The publisher does not have any control over and does not assume any responsibility for author or third-party websites of their content.

First Edition: November 2005

Library of Congress Cataloging-in-Publication Data
Staubach, Suzanne.
    Clay: the history and evolution of humankind's relationship with Earth's most primal element / by Suanne Staubach.
        p. cm.
    ISBN 0-425-20566-5
    1. Clay. 2. Clay–History. I. Title.

TA455.C55S73 2005
620.1'91–dc22

2005047154

PRINTED IN THE UNITED STATES OF AMERICA

10 9 8 7 6 5 4 3 2 1

For the truly wonderful and amazing
Olivia and Arielle
With Joy! And deep love.
Always.

And in memory of my cousin John Bergholtz

CONTENTS

ACKNOWLEDGMENTS   ix

INTRODUCTION   xi

1 COOKING POTS AND STORAGE JARS
Porridge, Ale, and International Commerce   I

2 HEARTH AND HOME
Ovens, Heat, and the Invention of Baking   19

**3 THE FIRST MACHINE AND THE DEVELOPMENT
OF AN INDUSTRY**   44

4 SET THE TABLE
From a Simple Bowl to a 2,200-Piece Dinner Set   64

5 A WORD OR TWO
The Invention of Writing and Books   97

6 THE MOST POPULAR BUILDING MATERIAL
Cities, Walls, and Floors of Mud   II2

7 SANITATION
A Nice Hot Bath, a Drink of Water, and Don't Forget to Flush   I40

CONTENTS

8 FARMING MADE EASY
Irrigation, Propagation, and Incubation 160

9 ELECTRICITY, TRANSPORTATION, AND ROCKET SCIENCE 180

10 TO YOUR HEALTH! 192

11 ART, TOYS, GODS, GODDESSES, AND FERTILITY 203

12 A FITTING DEATH
Urns, Gravestones, Companions, and Thieves 225

CONCLUSION 245

APPENDIX A: HOW TO MAKE YOUR OWN PINCH POT 248
APPENDIX B: MUSEUMS 251
NOTES 255
BIBLIOGRAPHY 260
CREDITS 271

ACKNOWLEDGMENTS

WRITING IS A notoriously solitary task, but somehow the process of writing this book has spilled out of my study into every other aspect of my life. I am deeply grateful to all who have been impacted and who have generously offered support and encouragement.

A special thank-you to my partner, Joe Szalay, who had to step around and over and in between piles of the books, papers, and index cards that accumulated throughout every room of the house during this project and who has been enormously, enormously helpful; to Gretchen Geromin, Dan Geromin, Aaron Geromin, and Az Geromin who have also encouraged me and urged me on, reminding me on occasion not to let other things get in the way; to all my bookselling friends and colleagues who have taken an interest, especially Marc Harnois and Julie Laurmark, who have ordered and received myriad books for me, and the rest of the bookselling crew at the UConn Co-op, Clare Morosky, Ian Schlein, Sharon Ristau, and Nikki Burnett, who have heard almost as much about clay from me as about books these past many months; and Jen Weinland, who has kept her eye on the project and hopefully will get her hands in mud again very soon; and my very good friends and cohorts in the ABA, ABFFE, IBC, NEBA, and NACS, especially Fran Keilty, Carole Horne, Mitchell Kaplan, Rusty Drugan, Nan Sorenson, and Kathy Anderson, who understand the book world all too well; and all the brave and hardworking booksellers throughout the country in whose hands the book now rests and for whom I have deep respect, affection, and sense of camaraderie; and thanks to reps Tim Allen and Karen Gudmundson, who have heard more about clay than they ever imagined possible yet kept smiling; and thanks to all my pottery friends, especially Maryon Attwood (not a potter at all really, but a crucial contributor to the clay world and a fellow projects conspirator and dear friend), Robbie Lobell, and Barbara Katz, and inspirations Karen Karnes, Ann Stannard, and Michael Zakin, and the potters of Clay Arts East and Centered; and thank you to Lary Bloom, who gave the earliest encouragement, advice and enthusiasm; and Samantha Staubach and Sydney Staubach, who have patiently waited for me to finish this book so I could fire their most recent pieces of pottery; and to my parents, Peg and Rich Staubach, who have a passionate devotion to both intellectual and hand work and who bestowed this gift of pleasure in the work of the mind and the hand to their children and grandchildren; and to Ed Knappman who found a home for this book; and Allison McCabe, who gave it an excellent one.

INTRODUCTION

Mud! Mud! Glorious mud!
Nothing quite like it for cooling the blood.
So, follow me, follow,
Down to the hollow,
And there let us wallow
In glorious mud.

—MICHAEL FLANDERS AND DONALD SWAN,
The Hippopotamus

YOU PROBABLY HAVE memories of playing with clay as a child, either some you dug yourself after a summer rainstorm, or perhaps clay your teacher gave you to squeeze and pinch into a small gift for your mother or father. I still remember scooping up the rich dark clay that rendered our backyard play space muddy and spreading it onto leaves to make myself "shoes." I have no recollection of what gave me the idea for such an enterprise, but I do remember the wonderful squishy, sticky feel on my bare feet.

Clay is ubiquitous. If you gathered it all up and spread it evenly over the surface of the earth like peanut butter, you would create a mud layer a mile in thickness. Certainly, there are places in the world where there is no clay—deserts, some mountain ranges—but in most areas of the world, it is readily available, often plentiful.

In addition to its abundance, two additional qualities have made clay instrumental to many key human endeavors: its plasticity and its durability after being treated with heat.

The word *clay* comes from the German word *kleben*, which means "to stick to." Wet clay is plastic in nature. If you squeeze it, it responds to the pressure and retains the shape you squeezed it into. Even more remarkable, once clay is fired ("baked" in a kiln), it becomes rocklike. If you drop it, it will shatter, but the pieces will last forever.

It is these three qualities of clay—its abundance, its plasticity, and its durability (even sun-baked clay has considerable durability)—that has made it so valuable to the progression of culture and the rise of civilization. Writing began on clay tablets; clay ovens and pots enabled the development of cookery; fired and unfired bricks made the building of houses and whole cities possible; and clay figures have had a vital role in religious practices and in the play of children. Today, clay is crucial to the computer and space industries, to biotechnology, to the publishing industry, for water clarification, and for a wide range of manufacturing processes. It is a part of your everyday life.

Your toilet is clay.

Your coffee mug is, too.

The coating on your favorite magazine's glossy paper is clay.

Some scientists believe that clay played a crucial role in the origins of life itself. They hypothesize that montmorillonite, a particularly fine-particled kind of clay, is the key to the transition between the nonliving and the living—that perhaps life began in a sort of muddy soup of clay and water, "energized" by a bolt of lightning. Indeed, the idea that the first humans were made of clay predates scientific thought. *Adam* is Hebrew for red earth, or clay.

Clay is alumina, silica, and chemically bonded water.

The ideal formula for "perfect" clay is $Al_2O_3 2SiO_2 2H_2O$. In reality, however, most clay varies somewhat and has impurities such as iron. The particles of clay are extremely small, generally 0.7 microns in diameter and 0.005 microns in thickness. A micron is $1/25,000^{th}$ of an inch. The particles are flat, two-dimensional, and electrically charged, and they touch on only two sides. When flooded with water, a strong attraction is created between the particles, yet the water also acts as a lubricant, allowing the particles to slide. This is what gives clay its plasticity.

Clay was once stone. Billions of years ago, as the molten magma of the earth slowly cooled, it crystallized into a hard crust, mostly granite. In time the granite was exposed to rain, ice, snow, wind, and glaciers. It expanded during the hot summers and contracted in the frigid winters. It was pummeled by sleet and hail. Floodwaters washed over it. The granite weathered and began to break apart. In some places it was exposed to organic acids from decaying vegetation and underwent chemical changes. After millions of years, both weathering and chemical action caused the granite to decompose, first into feldspathic minerals which, over time, released some of their minerals such as silica, calcia, potash, and soda, and then, as water molecules replaced the released minerals, the granite became clay.

Some clay remained in place where the original rock lay. This is called residual clay, and it tends to be "short" (less plastic). Other clay was carried away by wind and volcanic ash, by rains, rivers, and streams, and, in time, deposited great distances from the original bedrock. This is called sedimentary clay, and it is very plastic.

The great adobe houses of the American Southwest, the walled cities of ancient Mesopotamia, the cob houses of England are all built of this kind of "raw" clay—also called "unfired." Unfired clay is used to line ponds, to clean up toxic waste, as an ingredient in some cosmetics—spas even pamper their clients with mud facials. Much can be done with this unfired clay.

But it is firing that turns clay back to a stonelike material—and gives it most of its usefulness. Interestingly, the temperatures of a kiln (2,000°F to 2,300°F [1,100°C to 1,250°C] or somewhat higher) are the same temperatures as volcanic gases and molten magma are before they cool into rocks. The fiery heat of the kiln reenacts the great metamorphic processes that formed the world we know today.

Clay and the products created from clay are often classified by firing temperature.

Earthenware is "low fired," that is, to temperatures below 2,100°F

(1,150 °C). Above this point, the clay slumps and deforms. Commercial flowerpots and common building bricks are examples of earthenware. These clays can be red (terra cotta), white, gray, or black. Work made from this clay is not vitreous (waterproof) unless it is glazed (coated with a glassy material). Historically, earthenware was used for many utilitarian items (it was called redware in colonial America), and was coated with a clear lead-based glaze or, as in ancient Egypt and the southeastern United States, an alkaline glaze. The earliest fired pottery was earthenware.

Stoneware clays, which are fired at higher temperatures, are vitreous or almost vitreous even when left unglazed. They are very strong and relatively chip proof. Colors range from oranges and browns, to white, tan and gray. Early Americans made crocks and jugs of stoneware.

Porcelain, a dazzling white clay, free from impurities and vitreous, is fired to 2,300 °F (1,250 °C) or somewhat higher. Once fired, it is translucent and, when struck, has a clear sonorous ring. The main ingredient of porcelain is kaolin, which was first discovered by the Chinese in a mountainous region, Kao-ling, from which the word kaolin is derived (it means "high ridge"). Residual kaolin is found in South Carolina and Georgia in the United States; more plastic sedimentary kaolin is mined in Florida. Some of the largest beds of kaolin are found in England. All of these clays have very fine particles.

Ball clay is similar to kaolin except it has organic materials and is less pure. It was formed in ancient swamps rich in organic acids and gaseous compounds. Fireclay is similar to stoneware clay, but it has larger particles and is less plastic.

Shale is clay that is turning back to rock. In prehistoric times, it was the mud at the bottom of ancient lakes. Gradually, pressure has hardened it, making it cementlike. Shale can be mined and pulverized back into clay.

Earthenware, stoneware, ball clay, sedimentary, residual—no matter what the type, to most of us it is mud. It is the slippery mud along a riverbank, and the heavy mud in our gardens. It is the goo of our childhoods, and the mess our boots leave on the carpet.

I have had a love affair with clay for the past thirty years.

I've spent countless hours in my studio throwing spoon jars and wine cups and baking dishes on my kick wheel. I love the feel of the damp clay spinning in my hands. I like the long nights outdoors tending my kiln, and the yellow luminescence of the pots and the kiln interior when it is at last hot enough to shut down. I enjoy setting the dinner table with handmade dishes I have made myself or better, those that potter friends have made. It is exciting to me to realize that people have been doing these very same things for thousands of years. Even more exciting has been the discovery that clay, this most humble and common of earth materials, was the crucial component in many of the most important events of human progress.

COOKING POTS AND STORAGE JARS
Porridge, Ale, and International Commerce

My definition of Man is, "a Cooking Animal."
The beasts have memory, judgment, and all the faculties and passions
of our mind, in a certain degree.
But no beast is a cook.

–JAMES BOSWELL,
The Life of Samuel Johnson, L.L.D.

"THE FINE REPUTATION which Boston Baked Beans have gained," Fannie Farmer wrote, "has been attributed to the earthen bean-pot with small top and bulging sides in which they are supposed to be cooked."[1]

Farmer (1857–1915), author of the *Boston Cooking School Cookbook*, brought reliability to the kitchen when she invented level measurements. She was the doyenne of turn-of-the-century household cuisine, and her thoughts and advice on cooking were taken seriously. Many contemporary cooks agree with Fannie Farmer that the *only* way to prepare Boston Baked Beans is slowly in a traditional earthenware beanpot. The beans turn out tender and moist, and the flavors are nicely blended.

1

The industrious Puritan homemakers who originated this ever-popular recipe had brought with them the traditions of pease pottage (porridge) from England, and then adapted it to the ingredients on hand in their new country. They added maple syrup to their beans, and a bit of salt pork, later substituting molasses and mustard for the syrup when they became available. The pot was set close to the hearth, often nestled in a bed of warm ashes, on Saturday evening and left to "bake" all night. By this method, a housewife could keep the Sabbath (cooking was strictly forbidden on the Lord's Day in Puritan households) and still feed her hungry family a hearty Sunday meal.

Pease porridge, the national dish of seventeenth-century England and the forerunner of baked beans, was similarly made in an earthen pot left by the hearth. Dried beans, peas, lentils, or chickpeas plus whatever else was in the cupboard—a turnip, a few cabbage leaves, a scrawny rabbit—were tossed into the pot and slow-cooked. A family could, in essence, be served the same porridge day after day as the original porridge was stretched by adding a few more ingredients each evening. It was this practice that gave rise to the children's nursery rhyme:

> *Pease porridge hot.*
> *Pease porridge cold.*
> *Pease porridge in the pot nine days old.*

Ever since cooks have been using pots to make the family meal, they have taken those pots seriously, giving consideration to their proper shape, usage, and care. And they should—it was, after all, the invention of the first cooking pots and attendant storage jars, made of humble clay during the late Stone Age, that ensured the healthful well being of our species. These two vessels led to the development of the culinary arts, enabled security in times of drought, intensified international commerce, and gave us the gift of wine.

Scientists believe our ancestors domesticated fire about half a million years ago, after thousands of years of opportunistically enjoying the benefits of fires set by lightning. Our hunter-gatherer forebears acquired a taste for the occasional roasted animal and feasted on the charred remains of a fire's fatalities. They also indulged in the resulting baked roots and berries. In time, though, humans learned to make fire themselves, when and where they wanted it. Now they could enjoy warmth in winter, light at night, and commence a very rudimentary form of cooking.

Prior to the invention of cooking pots, cooking was primitive and often risky, though possible. In the Ukraine, the tusks of a woolly mammoth from the tenth millennium B.C.E. were discovered with their pointed ends stuck into the earth on either side of a hearth. Archaeologists suppose that the mammoth or parts of it were tied to a spit that spanned the tusks so that the meat could be roasted over the coals. Perhaps the tusks were reused to roast another animal long after the mammoth was consumed. Greens and nuts and a sort of "cake" made from ground cereals that were mixed with water to form a paste could be heated on rocks placed around a fire, or even on pieces of damp bark. Herodotus (c. 485–420s B.C.E.), the "father of history" and ethnography, described an ingenious potless method of cooking employed by the nomadic Scythians: "If they have no cauldron, they cast all the flesh into the victim's stomach, adding water thereto and make a fire beneath of the bones, which burn finely; the stomach easily holds the flesh when it is stripped from the bones thus the ox serves to cook itself." [2]

However, until Neolithic homemakers learned to make pottery vessels, real cooking could not begin, and many potential foods could not be eaten. The use of cooking pots meant that grains and cereals could be made into pottages. Greens, tubers, and tough chunks of meat could be conveniently simmered in water for soups and stews. Babies could be fed easily digested mush and weaned earlier, freeing their mothers for other chores. Legumes, roots, and pieces of fish could be combined to make nutritious dishes. New foods, such as cheese and butter, entered the diet. A meal's tastiness became almost as important as the subsistence value it offered, and cooks began to add onions, garlic, herbs, and other flavoring agents. A meal cooked in a pot offered more than the barest daily caloric requirement for survival; it offered pleasure, security, new skills, and an opportunity to socialize.

Clay figures of men, women, pregnant women, birds, animals, and sexless figures were sculpted, and often baked in the household fire for around twenty thousand years before vessels were widely made. Smooth, plastic clay was scooped from a nearby riverbank or clay bed and pinched and squeezed into various shapes. Many of these figures were probably used in religious ceremonies and fertility rites, or simply gave the maker delight.

But, remarkably, the oldest pottery vessels discovered to date were made by the Jomon peoples in Japan at least as early as the tenth millennium B.C.E., thousands of years before they began making clay figures. The Jomons lived along the beautiful seacoasts of Japan for ten thousand years. Jomon ware was used for cooking, food preparation, and storage. It was impressed with rope or shells or mats and had an allover cord design (Jōmon means "cord markings" in Japanese). The first Jomon pots had rather pointy bottoms, which were stuck into the soft sand or earth of the cooking fire. Later cooking pots had flatter bottoms. As the years passed, Jomon pottery became more and more elaborately constructed, with abstract and fantastical lizards, frogs, snakes, and curlicues on the asymmetrical rims, intricate designs impressed into the sides, and modeled images of animals. The unmistakable and highly creative style spread throughout Japan and, astonishingly, lasted until around 300 B.C.E.

The Jomons were hunter-gatherers. They lived in small communities of ten to twelve pit dwellings, which, after eight thousand years, grew to large villages, complete with cemeteries. In their later years, they practiced a limited form of agriculture and grew rice and millet. By 1000 B.C.E. they were making clay figurines, some with large eyes and oversized thighs, perhaps for some sort of religious ceremony. They made stone circles—again, perhaps for some sort of religious ceremony—and dressed in unisex outfits made of hemp. Some anthropologists believe that the Jomons were among the first wave of people to reach North America and postulate that, in the New World, they became the Sioux, Cherokee, and Blackfoot tribes.

Archaeologists are convinced that the Jomon potters were women. In fact, most archaeologists believe that Neolithic ceramics were largely if not exclusively the purview of women, who were in charge of the domestic family chores. Men hunted, built houses, and farmed. Women cared for the children and were responsible for food preparation and storage, clothing, and other close-to-home tasks. The work of pot making can be interrupted, and, in fact, damp, newly made pots must be set aside to dry, so the work was conducive to simultaneous child care. A mother could set her pot out in the sunshine to stiffen while she tended to her children. She could give her young ones pieces of clay to play with to keep them busy while she worked on new pots for the kitchen.

It appears that pot making sprung up independently in multiple areas throughout the world during the Neolithic era. Generally pottery skills developed when a culture embraced agriculture and became more sedentary, though the Jomons became potters two thousand years before agriculture was embraced anywhere. Jomon culture remained essentially a Mesolithic, or Middle Stone Age, culture except for pottery making, which anthropologists consider a Neolithic skill.

Handsome cooking pots and other utensils, decorated with bands and geometric designs, were made in Mesopotamia two thousand years after the Jomons began making their wares. The smooth, fine clay in the lowlands between the Tigris and Euphrates Rivers was especially workable and easy to shape. Soon afterward, the predynastic peoples along the Nile in Egypt were making pottery, as were the predecessors of the Greeks. Cooking pots were made by the indigenous peoples of Mexico and Central and South America, and in China and India and sub-Saharan Africa.

There are several theories explaining how people learned to make cooking pots and other vessels out of clay. One is that a prehistoric homemaker smeared the interior of a basket with clay to seal it so that she could prevent the seeds, nuts, and berries she wanted to save, from slipping through the open weave. Then she, or perhaps one of her children, left the mud-lined basket too near the fire and the reeds burned away. The clay remained intact and magically hardened in the fire. A pot! This likely explains the Jomons' invention of pottery since they had basketry skills, and the cord markings suggest reed basketry. A similar event could have taken place in various cultures throughout the late Stone Age world.

But not all Neolithic people made baskets. A second theory is that a fire pit might have been dug in naturally occurring clay, and realizing how much better such a pit held water once a fire had been burned in it than one dug in sand or dirt, Stone Age peoples might have begun to purposely line their pits with a coating of clay. The pit itself almost became a pot. The Celts, particularly the Celts of Ireland, were still lining cooking pits with clay as late as 500 B.C.E.

Or perhaps some clever and inventive ancestor simply thought up the idea. He or she made a little clay figure—a goddess, an antelope, a stylized lion—and tossed it into the evening fire at the mouth of the sleeping cave as an offering. Or a temperamental artist, in frustration and disappointment with how one of his or her small mud creatures turned out, threw it into a fire in disgust. The next morning, raking the ashes with a stick to stir the embers into flames, he or she discovered that the mud creatures had mysteriously become hard as stone. From this discovery came little bowls deliberately baked in the fire. It is likely that different peoples came to pottery in different ways.

Clay, being a plastic and forgiving material, lends itself to a number of forming processes, so the earliest potters, wherever they lived, were able to devise many ingenious methods for making cooking pots. A ball of clay could be beaten or pounded with the heel of the hand into a pancake or thin slab, which could then be pressed into a mold, such as a hollowed-out stone, or draped over a melon. Roman potters were fond of clay press molds and loved to use them to make pots with intricate impressed designs.

The early Sumerians made coils, or snakes, of clay that they wound into a circle and smoothed together. The more coils the potter added to the rim, the taller the pot. The Pueblo people made "corrugated ware" by leaving the outlines of the coils visible on the exterior of their gray cooking pots.

The potters of the Mimbres Valley in New Mexico used a paddle and anvil to beat their pots into shape. Some of the sub-Saharan potters did the same, as did the Chinese. In this method, the pot is started with a clay pancake or disk and built up with coils. Then it is shaped or thinned by holding a stone or wood anvil in one hand while beating the pot's sides against it with a paddle held in the other hand.

In many cultures methods were combined. Coiled pots could be built up on a base that had been molded or beaten into shape. A potter could pinch the pot into shape, starting with a ball in the palm of her hand and then pinching the clay between her thumb and index finger. She could finish with a rim made of a thin clay coil.

There was more work to pottery making than forming the pot, however. The clay needed to be dug and prepared. Debris and large stones would have to be removed, and some sort of temper, such as sand, would be added to make the pot sturdy enough to withstand the rigors, especially the thermal shock, of the cooking hearth. And, to be useful, pots had to be fired. This could be accomplished in a simple bonfire of gathered sticks and brush. Or a pit could be dug and the pots could be placed on a bed of straw at the bottom of the pit and covered with thin pieces of wood or dung. Firings were of short duration and left the pots porous. This porosity, together with the sand temper, enhanced the durability of the cooking pot, so that it would last through the repeated heating and cooling cycles of culinary use.

■

WILD CEREALS AND grains flourished after the last glacier receded and the earth's temperatures warmed. They were among the initial crops cultivated by Neolithic farmers. But grains and cereals cannot be eaten and digested directly from the stalk; they must be cooked.

With the advent of cooking pots, cereals could be boiled in water, or even goat's milk, to make nourishing gruel or pottage, the world's first comfort food. Indeed, the word *pottage* means "from or in a pot."

Pottage became a universal staple and, like pottery, appears to have been independently invented by each culture, using whatever grain or cereal was grown locally. Native Americans made a maize porridge, which the colonists adopted and called corn meal mush. The Romans ate barley porridge and used millet porridge for their polenta (though by the sixteenth century their Italian descendants switched to corn meal mush for polenta). Millet porridge became a staple in Africa. Adventurous cooks, like the Carthaginians of the sixth century B.C.E., mixed an assortment of grains for porridge. Porridge was so important to the European diet of the Dark Ages that during the desperate grain and cereal shortages of this time, Christian monks added clay to it to make it go further.

Cato, the Roman writer, statesman, and soldier, gives us this recipe for a hearty pottage: "Add a pound of flour to water and boil it well. Pour it into a clean tub, adding three pounds of fresh cheese, half a pound of honey, and an egg. Stir well and cook in a new pot."[3] He meant a new pot, too. Romans like himself, who considered terra-cotta cooking pots a throwaway item, kept their local potters in business.

Today, many people begin their day with a serving of cold cereal. They shake some commercially produced flakes made of wheat, oats, or corn into a bowl and add cold milk, sugar, and perhaps some fruit. Others stir boiling water into some "instant" or "quick" oatmeal. Ironically, though huge corporations make these cereals for us with crops grown on vast corporate farms, and we purchase them ready-prepared in boxes from the grocery store, the breakfasts that so many of us eat are not modern at all, but merely a version of late Stone Age cuisine!

As the Mesopotamians soon learned, the trouble with a pot of porridge is that it doesn't keep well. In fact, depending upon the cereal used, it ferments, particularly barley. And if you add more water to fermented barley, you get beer or, more properly, ale, a beverage the ancients became very adept at producing (beer as we know it was not produced until the end of the Middle Ages in the sixteenth century). The invention of pottery vessels led not only to delicious and healthful diets, but also to the brewing and consumption of alcohol.

The Egyptians believed that it was the god of agriculture, Osiris, who invented ale. He had made a soupy porridge of sprouted barley mixed with the sacred waters of the Nile. But, being a busy god with many demands on his time, he didn't eat it right away, and absentmindedly left it out in the sun while he attended to other more urgent matters. When he came back to take a bite, the brew was bubbly. It had fermented. Cautiously—and bravely—he took a sip. "Delicious!" he shouted and drank the whole potful.

Despite the Egyptians' claims for Osiris, the Sumerians and the Babylonians were making and drinking ale before the god of agriculture had his first indulgence. Sumerian women made eight kinds of beer from barley, eight from wheat, and three from a mixture of grains. They brewed and sold their alcohol from their homes under the guidance of the goddess Ninkasi, "the lady who fills the mouth," though the process was so unreliable that Hammurabi railed against the frequent poor quality and high prices in his famous Code.

The Greeks believed that Dionysus, the god of wine, left Mesopotamia because the people who lived there were drunks. Perhaps he merely jealously disapproved of ale, or perhaps there was some truth to his judgment. In the *Epic of Gilgamesh*, the Mesopotamian account of the Flood, or Great Deluge, the Sumerian king, Utnapishtm, tells Gilgamesh that after his workmen finished building the ark, he gave them "ale and beer to drink, oil and wine as if they were river water."[4] The biblical version of the Flood story bluntly refers to Noah's drunkenness after the waters abated.

Literature aside, we know that a common Sumerian man who worked in one of the temples was allotted almost two liters of ale a day. An administrator was given more than eight (one hopes he used it as currency or for barter and did not feel compelled to drink his entire allotment each day). Whether or not the Sumerians drank too much ale, it was probably healthier than the sour drinking water that came from their irrigation canals. It was these extensive canals that, after years of use, increased the acidity of the agricultural lands of the Sumerians and greatly diminished the grain crops. With such shortages, the Sumerians were forced to cease making ale. However, not willing to give up their pleasure, they took up drinking date wine, which they made from the fruit of the trees that they planted along the riverbanks rather than in their irrigated fields.

The people of the Nile, however, continued to brew ale. Around 1500 B.C.E. they realized that old, well-used pottery jars more reliably produced good ale than new jars. Used jars, worn with chips and cracks, with the residue left in the pores from previous brews, fostered the growth of the natural yeasts necessary for the fermentation process. Across the Mediterranean, the sociable Athenians prized this predictable and tasty Egyptian ale, and imported it in great quantities.

Although the most common Egyptian ale, *haq*, made from the red barley of the Nile, was very low in alcohol content, the rest of the ales the Egyptians produced probably had a 12 percent alcohol content. One Egyptian writer warned, "Do not get drunk in the taverns in which they drink ale, for fear that people repeat words that may have gone out of your mouth, without your being aware of having uttered them."[5]

The taste for ale spread from Greece to the eastern Adriatic, Gaul, Spain, and Germania. The Romans, however, never liked it. They preferred white wine made from grapes.

Vitis vinifera, the wild grape vine, and parent of more than three thousand contemporary varieties, may have originated in the Caucasus, but was growing throughout the Mediterranean prior to the Neolithic era. Stories of how the first wine was made are similar to the stories of how the first beer was made: a pot of drying grapes was left out and forgotten by an ancient housewife; when she returned, she tasted the fermented juice, the first wine, and liked it. Viticulture and wine making quickly spread.

Egyptians continued to drink ale, and reserved their wine for use in their temples. Nevertheless, Egyptian vintners became highly skilled. They crushed and squeezed the harvested grapes in large cloth bags. The wine itself was stored in tall clay jars, amphorae, which were carefully labeled with all the sorts of information we find on a modern wine label: the date when the wine was made, the type and color of the wine, the name of the vineyard and the grower, and, once a lively trade developed, the name of the wine merchant.

The Greeks imported huge quantities of Egyptian wine as well as wine from Palestine and Mount Lebanon. Soon they grew their own vines on rocky hillsides. Greek vintners artfully blended wines from different vineyards and different grapes in their amphorae to achieve the best taste. Connoisseurs now favored these hearty Greek wines and Greek merchants grew rich exporting them.

The Romans appreciated fine Greek wines but by the end of the second century B.C.E. their own vintages had surpassed the imports in production, popularity, and quality. A well-tended Roman acre could produce sixteen hundred gallons. Slaves were engaged in the crushing process and the juice was strained through baskets into *dolia*, special earthenware fermenting jars. The interiors of these jars were coated with pine pitch and the wines were left to age in them a minimum of two to three years, and often fifteen to twenty-five years. Like Egyptian and Greek wines, Roman wines were thick, and intensely flavored. They had to be mixed with water, in specially designed mixing bowls, called *kraters*, before drinking.

"It is unlikely," writes food historian Reay Tannahill, "that wine was made regularly until pottery was invented, since its making and storage would take up a disproportionate number of containers."[6] In fact, it would be nearly impossible to make wine or ale without fermentation and storage jars.

Ceramic storage jars not only enabled the production of alcoholic beverages, they were the shipping containers, closets, and pantries of the ancient world. They provided insurance against a poor crop. Egyptians kept an inventory of grains in huge jars so that if the Nile failed to flood, they could still eat. The cellars of the Palace of Knossos on the isle of Crete were stocked with *pithoi*, storage jars, six feet tall, set in magazines and filled with honey, wine, and other foods. Houses in Egypt and Mesopotamia were outfitted with similar storage jars, often in the cellar or kitchen, filled with grain, oil, beer, and sometimes wine. There is evidence that they were used to store clothing. Centuries later, the Essenes used jars to protect the scrolls they hid in desert caves. Romans kept amphorae filled with wine on their rooftops.

Elsewhere in the world, and down through the millennia, storage jars played a crucial role in the daily lives of citizens. Louise Allison Cort, the curator for ceramics at the Smithsonian's Freer Gallery of Art and Arthur M. Sackler Gallery, says, "Throughout Asia, until recently, stoneware storage jars were indispensable for the countless needs of domestic life, commercial transactions, and long-distance trade. (It is easy to forget that stoneware jars were equally indispensable for many tasks of life in the United States.) Jars held liquids or foodstuffs—water, oil, wine, rice beer, soy sauce, pepper paste, vinegar, fish sauce, pickled vegetables, and salted fruits. They protected seed grain, medicinal herbs, and textiles from humidity, insects, and rodents. They concealed stashes of coins or jewelry. They fermented indigo baths for dyers and served as sounding devices beneath Japanese theater stages. They interred sacred relics in the foundations of Buddhist pagodas, preserved sacred texts in preparation for the anticipated end of the world, and harbored the cremated remains of the dead. Jars were more precious than gold when negotiating marriages or peaceful conclusions of inter-village disputes.

"Large utilitarian jars stood in kitchens, storerooms, or barns, outside dwellings beneath the eaves, or in the open fields."[7]

In 1350, the Arab traveler Ibn Battuta wrote of martaban, the famous large jars made and named for the port city of Martaban, Burma, that they were "filled with pepper, citron, mango all prepared with salt, as for a sea voyage." Later, in the sixteenth century, Duarte Barbosa, a Portuguese adventurer, said, "In this town of

Turkish salt jar, earthenware. One of the oldest and most elegant of shapes. Thrown on a potter's wheel.

Martaban very large and beautiful porcelain vases are made, and some of glazed earthenware, of a black colour, which are highly valued amongst the Moors, and they export them as merchandise ... " Martaban were popular in the Middle East and India too, and were reputed to be able to undertake great journeys on their own, marry, and even have children.

Storage jars were critical to the well-being of diverse populations. The potters of the Harappan civilization, which flourished almost five thousand years ago in the Indus River Valley and most of northern India, relied on their beautiful dark red storage jars, which they decorated with leaves and vines and peacocks painted in black. The Incas, whose empire in 1450–1550, encompassed all of the inhabited Andes Mountains, carried large, narrow-necked jars of water on their backs. Because earthenware jars are porous, they sweat. The ensuing evaporation cools the water inside, a principle understood, at least empirically, by most ancients.

Perhaps the greatest importance of storage jars was that, being excellent shipping containers in a world that relied heavily on sea trade, they opened up new opportunities for international commerce (and the concomitant exchange of ideas). Historian Maguelonne Toussaint-Samat, in discussing the olive oil trade in her book, *History of Food*, writes, "Great merchants came from such oil-producing countries as Phoenicia, Crete, and Egypt to the Mediterranean basin and even farther from the sixth century B.C. onwards, the Scythians of the southern steppes of Russia came to stock up with oil at the prosperous Greek trading posts of the Black Sea which later became the spas of the Romans. Depositories of oil jars, such as those of Komo in Crete are evidence of the importance of this trade."[8]

The Greek economy was heavily dependent upon Mediterranean sea trade. Greek merchants used large ovoid jars, tapered into a narrow but thick foot, as shipping containers in their lucrative wine export business.

The Romans shipped wine, olive oil, fish sauce (garum), and dry goods in amphorae to distant ports throughout their empire. Ships traveled to North Africa, Gaul, Nubia, and Britain. In 250 B.C.E. the Roman jars were similar to the Greek ones, but gradually they developed a variety of shapes, usually with handles or lugs on the shoulders for attaching carrying-ropes. These commercial containers were either stamped with the name of the producer and sometimes the destination and the contents or the information was painted on the side of the jar. Many intact amphorae have been recovered from shipwrecks and undoubtedly more await discovery. The early potters would have been pleased to learn that though a ship sank to the ocean floor in a gale or storm, the clay jars remained whole, and continued to hold their contents safe for thousands of years until they were discovered by twentieth- and twenty-first-century archaeologists and divers.

As civilization progressed, new types of cooking pots were created: pie pans, tart pans, stew pots, pudding pans, and molds. British country potters made panchions, ample bowls with straight sides that flared at the rim, rather like an oversized deep pie pan. Panchions were used for baking and the biggest ones were used for washing. British country potters also made specially shaped ceramic dishes for cooking fish (fish shaped, naturally) and for ham (yes, shaped like a ham!). And they made rectangular baking dishes with dividers, just in case you didn't want to get your food mixed up.

Eventually, new materials were utilized for cooking pots and for storage. The Chinese clay cauldron used eight thousand years ago was replaced by a bronze version. Copper and iron came into use, followed by stainless steel and plastic. But clay pots and jars have not disappeared from the household.

The modern Chinese cook favors a sand pot for stews and casseroles. Made of earthenware heavily tempered with sand, these pots usually have a domed lid and a handle on the side and, for safety, are sometimes encased in a wire cage. The Moroccans developed the *tangine slaoui*, made of glazed terra cotta. The base is a round, shallow dish. The lid is conical, like an upside-down funnel. Set over a brazier, tangines are used for slow-cooking stews such as mutton with quinces and honey.

Cooks in many countries rely on a shallow, lidless clay pot, round, sometimes oval, for baking or braising. The Spanish call theirs a *cazuela*. The Provençal version is a *tian*.

The French, who have not been been able to agree about the

proper ingredients (lamb? sausage? pigs' feet?) for cassoulet, a traditional rustic entrée, do agree that this bean dish is only to be cooked in the earthen pot of the same name. It is a tall narrow pot, rounded at the bottom and narrowed at the top. It is similar to the bean pot Fannie Farmer extolled and to the redware "bulge pots" or stew pots made in New England, particularly colonial Connecticut.

This ability to combine and mix ingredients in many ways is the great benefit of the invention of cooking pots. During the Middle Ages, vertical and horizontal *composita*, or compositions, were made. A vertical composita, in which the ingredients were stacked on end, necessitated a taller pot than a horizontal composita, in which the ingredients were layered. An early American version of layered food, descended from a meatless European version served during Lent, was the Pennsylvania Dutch *Gumbis*. This consisted of cabbage, apples or pears, perhaps venison, caraway seeds, maybe lard, and kale, all layered in a clay pot.

Today's casseroles are usually cooked in a deep round pot with straight sides and a lid, called (what else?) a casserole. The word comes from the ancient Greek *kuáthin*, for "cup." This evolved into *cattia*, late Latin for "pan" or "basin." The French then got into the act and it became *casa* in Provençal, and *casse* in Old French. What the French were cooking was a rice dish molded in the shape of the pot in which it was cooked and filled with chicken or other ingredients the cook had about the kitchen. The original English casserole was also a rice dish until the 1870s when it somehow lost the rice and acquired meat and vegetables. Our typical contemporary casseroles might have noodles, pasta, canned soup, tuna, vegetables, sauce, or leftovers from the refrigerator.

Ceramic casseroles are widely available commercially in many shapes and sizes. They are also a favorite item for studio potters to make.

He didn't call it a casserole, but even Hitler embraced the one-pot meal, called the *Eintopf*. He ordered the Germans to eat a one-pot meal one Sunday a month from October until March. They were to give the money they saved by eating this frugal fare to a winter relief fund for the impoverished.

Now you can go to a gourmet shop or browse a mail-order catalog and find many specialty ceramic cooking pots. There are little covered pots for baking garlic, pots that are really large closed clay jars turned sideways and sliced to form a lid and a bottom used for cooking chicken, rectangular clay pans for bread, and clay disks for crusty pizza. There are custard cups, ramekins, and small one-handled pots for making onion soup. And, yes, there are bean pots.

Except for specially formulated pots, or well-tempered earthenware, today's clay cooking pots generally can't go directly onto the burner of a modern stove without risking breakage. Our stove burners, beginning with Ben Franklin's invention of the iron "Franklin stove," reach temperatures that are too high, and that can change too fast, for most modern ceramics to withstand.

But throughout history, cooks have used clay cooking pots nestled in a bed of ashes or coals, covered with coals, placed in front of the flames and moved about, suspended over the fire, or, once they were invented, set in clay ovens. They were even, for a while, submerged in water in iron cauldrons so that the food inside could be steamed. Moroccans actually make many of their braziers *naffekh* of earthenware, and set a pot on top.

Three-legged cooking pot, burnished smooth and low fired in a reducing atmosphere from Ethiopia.

In *America Eats*, William Woys Weaver tells us that despite the widespread dissemination of iron cookstoves and iron cookware, with the Arts and Crafts movement many Americans reverted to clay cookware. He writes, "In Keene, New Hampshire, the Taft Pottery brought out a patent earthenware saucepan for use on the cook-stove. Its design was a revival of the old colonial *pipkin*, in this case a pipkin without feet. In fact, cooking in stoneware grew so much in popularity by the end of the 1890s that it was even called a 'renaissance,' in spite of the prevalence of iron cookware at the time."[9]

During the 1970s, studio potters Richard Behrens, Karen Karnes, Michael Zakin, M. C. Richards, and Ron Propst, among others, experimented with making "flameware" and developed formulas for high-fired clay bodies that could sit on a burner or even in a fireplace. Karnes is still making beautiful flameware today. The Corning Glassware Company developed a "high-tech" ceramic cookware that could go straight from a cold refrigerator to a hot oven without breaking.

International merchants no longer use large amphorae as shipping containers and wine is now made in stainless steel vats, but ceramic storage jars continue to play a role in daily life. Ancient civilizations developed many jars shaped for specific uses: to hold perfume, medicines, jewelry, or small amounts of particular foods. Households of the Middle Ages used cream pots and butter pots and salt pigs (an open-mouthed salt cellar) for storing their staples. The nineteenth-century Portuguese traders shipped ginger beer in stoneware bottles. Early Americans kept pickled tomatoes, cucumbers, and eggs, sausages, crackers, flour, and sugar in their ceramic storage jars.

In the dimly lit and cool pantries of southern Anatolia, large clay jars, with a piece of embroidered cloth draped between the mouth of each jar and its lid, are still used to hold staples such as "sugar, bulgar (cracked wheat), vermicelli, chickpeas, lentils, beans, fruit and vegetables, as well as jams and pickles, grape molasses, fried or roasted meat, mince meat, . . . salted fish, almonds, walnuts and other tidbits. The spaces between the jars are filled with onions and potatoes,"[10] writes Nevin Halici, a Turkish chef and author. In Morocco, flour and oil are stored in "massive 'Ali Baba' jars, or *khabia*," and "amphora [are] packed with *tangia*, the comforting 'bachelor's

Salt-glazed ginger beer bottle from Portugal.

stew,' and garlicky *hergma*, a robust sheep's trotter *tajine*,"[II] Jill Tilsley-Benham writes in her essay in *The Cook's Room.* In Crete, pithoi, or large earthenware oil jars and amphorae, are still handmade in the same way as during the reign of King Knossos. However, they are no longer required for the king's pantry and instead they are exported to Great Britain and other flower-conscious countries where gardeners use them as bold accents and focal points in their landscapes. In the modern American kitchen you might find a jar crammed with wooden spoons, spatulas, and other cooking utensils; individual ceramic "canisters" for flour, sugar, tea, and coffee lined up on the countertop; a fancy cookie jar; and (a recent innovation) a compost crock for storing kitchen waste until the cook has time to go to the outdoor compost pile. The desire for "progress" seems to be genetically encoded in our species. Once cooking pots and storage jars were made, potters began working to improve them. With these improvements came new sorts of pots, like bottles and jugs, and fancy baking dishes and molds. And, in seeking better ways to make their jars and cooking pots, potters made, as we shall see, great technological advances in how they worked, which led to more technological advances in other fields of endeavor.

HEARTH AND HOME
Ovens, Heat, and the Invention of Baking

Nothin' says lovin' like somethin' from the oven.
—PILLSBURY COMMERCIAL

Keep the home fires burning.
—LENA GUILBERT FORD

CLAY WAS CRUCIAL to the development and manufacture of cooking pots and storage jars, and hence hearty and nutritious family meals—plus it was the ideal material with which to build an oven.

Clay ovens offer far more culinary possibilities than a stone hearth. With an oven, you can bake.

At surprisingly early dates, enterprising cooks used local clay to build ovens. Archaeologists have discovered numerous remains of bake ovens throughout the ancient world—in Jarmo, Iraq, dating from 5000 B.C.E., and in Tepe Sialk, Iran, dating from 3300 B.C.E. Fifty-six bread ovens dating from about 1000 B.C.E. have been excavated in the Jordan Valley. Large circular ovens (three and a half feet high and slightly wider) were found in Mohenjo-daro, in the Indus

Valley. Single-chambered beehive ovens have been discovered along the eastern Mediterranean dating from ten thousand years ago.

The Sumerians, who lived in cities along the Euphrates by 3000 B.C.E., developed large bakeries as part of their temple economy.[1] The kitchens in their temples were equipped with "large clay beehive ovens."[2] The *tannur*, a tall, beehive-shaped clay oven, is frequently mentioned in the Hebrew Bible. The tannur was open at the crown and outfitted with an air vent in the base for the fire. A cooking pot was sometimes placed on the top of the tannur, but primarily the ancient Hebrews used it to bake their flatbread. The baker slapped the sticky dough against the hot wall, where it stuck fast until the moisture was baked out. Then, scarcely a minute later, just as the evening meal was about to drop into the coals, the baker deftly plucked it from the hot oven. Similar unleavened breads or flatbreads (*chapati* in Hindustani; *nan* in Persian, Urdu, and Punjabi; *pita* in English [from the Greek] have been an important staple of the human diet for thousands of years.

Tannurs were fueled with charcoal, wood, dry patties made of dung, chopped straw, or any other combustible that was readily available to the household. The tannur (which has various related spellings such as tenur, tenner, and tandoor) spread from Persia to northern India, where the construction and use of tandoors was perfected. You can see the splendid results in most Indian restaurants today.

The tandoor oven is made from clay tempered with grass and rolled out into slabs. The slabs are stacked and smoothed together to form a cylinder. The lip curves in slightly. In traditional Indian households, the tandoor is set into the ground. In modern Indian restaurants, the tandoor is usually encased in a stainless steel jacket. Today, "authentic" tandoor ovens are manufactured in India and Great Britain and exported all over the world.

Tandoori cuisine was refined in nineteenth-century Peshawar to include bits of meat on skewers as well as flatbreads. Aficionados of the tandoor believe the clay of the oven adds to the rich, slightly smoky flavor of the foods.

Interestingly, many Indian women, especially in the villages, also use a stove made of mud called a *chula* or *sigri*. A chula can be square or round with thick walls. It has four raised clay knobs on the top, on which to set the cooking pot. Below the pot is a round opening to the fire. There is another opening at the base of the chula, in the front for stoking. Local potters make the mud stoves of clay kneaded with horse dung. They are not fired, but with use, they become fired. Typically, a chula lasts for a year before it has to be replaced with a new one.

Mud stoves common in India. There are "two burner" and "one burner" styles.

Flatbreads baked in a tandoor are lighter in texture and better tasting than flatbreads cooked on a stone or pan in an open fire, yet this is not the bake oven's major contribution to the culinary arts.

Its major contribution is that it made baking *raised* bread possible.

To bake leavened bread, or bread that rises, three things are needed: yeast (or another leavening agent), wheat or rye flour, and an oven. As H. E. Jacob explained in *Six Thousand Years of Bread*, "Bread is a product baked in a properly constructed oven and from a dough that has been raised by yeast or some other leavening agent. Some of the gases produced by the leavening agent are imprisoned in the dough. The pores containing these gases are hardened and made permanent by heat. Only dough made of wheat or rye possesses the ability to retain the gases; this is due to the specific properties of the proteins peculiar to these grains."[3]

The Egyptians possessed all three prerequisites. They were an agrarian society and grew fields of wheat. They had yeast, from the barm left over from the fermentation of their beer. They possessed excellent ovens. In addition, they tended to be highly innovative. The earliest Egyptian ovens were probably made of a few flat stones stacked together. Later (around 2500 B.C.E.), pots were turned upside down over a bed of embers and heated. The hot pots were then filled with whatever was going to be baked, and placed rim to rim. The food was baked in the residual heat.

By 2000 B.C.E., the Egyptians had tall, conical-shaped clay ovens, which they loaded from the top. Inside was a grate; the ovens were heated with a fire, and when the fire had died down, bread was placed inside and the top of the oven was closed with a clay disk. Leavened breads took longer to bake than flatbreads. Sticking them to the sides of the oven was impractical. Instead, they were baked in clay molds that were set on the interior grate. Bakers became very creative and braided their dough or twisted it into pretty shapes. They added poppy seeds, sesame seeds, eggs, honey, nuts, dates and whatever else captured their imagination and soon there were more than fifty different kinds of bread coming from the advanced clay ovens along the Nile. Cooks discovered that if they saved a piece of "starter dough" from the previous batch of bread and used it in the next batch, the new bread too would rise. Today, we call this sourdough.

The rounded loaves of leavened bread, crusty on the outside and soft on the inside that the Egyptians baked in their lidded clay ovens became their favorite meal. They loved their bread so much that throughout the Mediterranean region they became known as "bread eaters." Herodotus scoffed that they "kneaded bread with their feet and clay with their hands."[4] However, travelers who tasted the airy Egyptian bread also found it irresistible.

Nearly every Egyptian household boasted at least one terra-cotta oven in the courtyard. "In larger houses," Edda Bresciani writes in *A Culinary History of Food*, "it was possible to find a whole room devoted to cooking as in the house known as 'the house with three ovens' from the period of Thutmose IV, discovered in Gurna by a University of Pisa expedition. This room had a *canun*, a cooking range with three fires and a surface for pans, as well as a place to put the water jar (a hole carved out of the rock floor with a drainage channel). Clay jars with openings near the base, used to store legumes, cereals, spices and condiments, stood on the floor."[5] Some ovens were placed on the flat rooftop. Most were outdoors, under a lattice roof.

Egyptian artists depicted bakers at work in great detail. We can see the bowls to mix the dough, the molds, and the ovens. Illustrations of bread baking covered the walls of tombs, left as instructions for the afterlife.

Archaeologists have also found evidence of domed beehive ovens in Turkey and eastern and central Europe. The peoples of the Neolithic Sesklo culture, which flourished in Thessaly and southern Macedonia from 6500 to 5600 B.C.E., built large domed ovens in their courtyards. They built attached clay benches along one side of the ovens, or circular raised platforms in front. Here, they could sit and prepare the evening meal or work on some other household task. Massive clay ovens have been discovered inside the remains of Neolithic houses in Central Bulgaria. And the peoples of the Butmir culture, which inhabited Bosnia, near modern-day Sarajevo, from 5300 to 4200 B.C.E., built beehive bread ovens inside their houses. Their ovens had stone foundations covered with a thick clay floor. The Butmir ovens had fired clay cylinders embedded in the thick walls of the dome, probably to vent some of the heat out of the ovens to provide warmth on a chilly evening.

Indeed, peasant families throughout the cold climates of northern, central, and eastern Europe relied on clay ovens for heating as well as cooking. Their enormous ovens, or stoves, dominated the interior space of their houses. Families would hover close to the warm clay walls, sleep on attached clay benches, or even sleep on *top* of the stove! The household cook would roast meat in the fire and, when the fire had died down, bake bread in the residual heat. Clothes, damp from a day spent out working in the rain or snow, or freshly washed, were draped nearby to dry. A family's livestock was allowed to bask in the warmth of the stove, fattening up in comfort before being slaughtered and cooked in the very flames that offered them winter protection. The daily life of European peasants was centered on the oven.

The design of early beehive ovens remained relatively unchanged for thousands of years. They might be whitewashed, there might be niches in the walls, but the domed shape and thick walls continued for millennia.

In the parts of the world where the winters were cold and the oven was built inside the house and used for warmth as well as for baking, smoke and soot hung in the air and sparks and fire were a constant threat. You could cook your winter meals indoors, protected from the frigid outdoor temperatures, but, while the fire burned, your eyes would sting and you would likely suffer a runny nose and headache. Fortunately, because thick clay walls retain heat, and slowly give it off, a fire would not have to burn constantly.

Still, peasants sought solutions to these problems and inconveniences. In Sweden, as a precaution against fire, the roofs of houses were covered with a layer of turf. In Germany, peasants constructed large clay vaults over their stoves to act as spark catchers. In Latvia they wove loose twig baskets, which they covered with clay and inverted over their stoves. Where houses were not built of clay, the interior walls close to the oven were coated with a smooth layer of fireproof mud.

In the high altitudes of the Alpine regions of old Europe, stove builders, probably potters, inserted earthenware pots into the thick clay walls of their stoves during the building process. The pots were set with their rounded bottoms facing out and the mouths facing into the stove and would have allowed heat to leave the stove more quickly. Though one of the great benefits of a thick-walled masonry oven is the heat retention and the length of time the stove radiates a gentle heat, in a very cold climate there are going to be nights when you want quick hot heat to warm a room. As a solution, Alpine potters also placed a few pots in the reverse position, with the mouths opening into the room and the bottoms inserted in the oven walls. These were called "fist warmers." If your hands were cold and stiff from working outdoors, you could thrust them into the cozy interior of the embedded vessel for relief.

Around the Mediterranean where the climate was mild, outdoor beehive ovens continued to be popular down through the years. Spanish explorers spread to the New World, the use of the adobe *horno*, their version of the beehive oven, and the *chiminea*, a clay heater shaped like a giant bottle with a hole cut in the front. Both are still in use in South America and the southwestern part of the United States today. In fact, the chiminea, though actually unsuitable for cold or rainy climates, has become recently fashionable on the patios of Europe, New England, and Canada. Originally used to take the chill out of the night air, the chiminea was kept indoors near a window and laid with a very small fire. In the morning the embers and remaining warmth were used for cooking.

Simple domed beehive ovens built of the most humble and inexpensive materials—local mud or clay—are still in use throughout the world today. Skilled oven builders in Iran use clay to build ovens seven feet in diameter.[6] Outdoor clay beehive bread ovens are popular in Canada, the Middle East, Italy, and France. The reason this very simple oven has enjoyed continuous use since the Neolithic era is that it is relatively easy to build also, it works beautifully. The thermal mass of a clay oven not only retains heat, it offers three kinds of heat: the radiant heat from the thick walls, the conducted heat from the hot floor where the fire burned, and finally convection heat, created when the moisture from the baking bread dough escapes and sets the air in motion. These three types of heat explain why bread from a clay oven is tastier than bread baked in a modern steel oven.

You can build your own beehive oven and bake bread for your family, just as our ancestors did thousands of years ago, if you don't mind getting a little dirty. It is a project that you can do alone, but it is easier with a few indulgent friends (which is generally how these clay ovens have always been built). To begin, you must clear all the grass, brush, leaves, and sticks from an outdoor area about 100 to 150 square feet. The clearing can be round or square.

Gather an armload of flexible saplings (willow is excellent), two or three wheelbarrow loads of sticky clay that you have dug nearby (or purchased), a half a bale of chopped straw, and, if you wish, a pail full of sand. Dump the clay onto some of the cleared ground (not in the center where you will be building your oven). Mix the clay with the chopped straw and the optional sand.

Scratch a circle in the dirt in the middle of your clearing. You can do this freehand, or you can push one end of a sapling into the center of the cleared area and tie to it a length of string equal to the radius of the oven you wish to build. I suggest a foot and a half. Affix a short straight twig to the other end, and use this twig to draw the circle. Cover this circle with a six-inch layer of clay. Pat it smooth.

Now push one end of a sapling into the circle perimeter and bend it over until it touches the opposite side. Stick that end into the earth too. Continue this all the way around the circle. Make sure the saplings are securely implanted in the earth.

Carefully weave crosspieces so that you have a fairly open basket. Once you have your stick basket igloo finished, cover it with the rest of the clay and straw mixture. To make your walls substantial enough you will want to add a layer and let it dry for a day before adding another layer. Do this until you have a nice thick wall. Leave a small, arch-shaped opening at the bottom. If you wish, you can use your empty pail as a mold for the opening, and remove it before you fire the oven. Our ancestors would not have used a galvanized pail, but they could have used an old potsherd (which you are unlikely to have lying around ...). Let your oven dry thoroughly. Depending upon the weather, this will take a few days or a few weeks. If it rains, be sure to cover the oven to protect it. Once it is hard to the touch both inside and outside, light a very small fire to completely dry it out.

Now you are ready to bake.

Mix up some bread dough and set it aside. Make a small fire of dry kindling wood inside your oven. Keep adding sticks until it is good and hot. The sapling frame will burn away, which is okay (in some of the ovens that the anthropologist Marija Gimbutas excavated, she could see the indentations from the burned-out saplings in the oven interiors.) Once the fire has died down, rake out the coals and ashes. Now you can put your bread inside to bake.

Don't be alarmed if your first loaf is not perfect. Baking in one of these ovens takes a bit of practice.

What I have just described is one of the oldest and simplest methods of building a clay oven, but of course, there are many variations. The horno is built of sun-dried adobe bricks. The bricks are laid and mortared with wet clay in a circle. When the walls are high enough, the bricks are either corbelled or laid at a tilt to form the dome. The horno has a hole opposite the door to create a bit of a draft. While baking, both the door and hole are plugged with rags or wood boards sealed to the horno with clay. Alternatively, a dome form can be made of damp sand, rather than saplings. Once the clay oven is dry, scoop the sand out with your hands.

Ovens can be built on a base of stones and firebricks. In parts of the world where the ground freezes, they are built on foundations with footings below the frost line. In modern urban or suburban settings, or in snowy climates, they are often built of fired brick. Doors are sometimes made of cast iron. The mud oven might have a lime plaster coating or be built under a sheet metal roof.

Whatever the construction variations, the ovens themselves remain essentially the same. And they are extremely effective.

■

DOMED CLAY OVENS, though nearly universal, were not the only type of stove, range, or heating device made of clay, however. An ancient combination cooking and heating system found in the rugged terrain and harsh climate of Afghanistan is the *tawakhaneh*, or hot room.

Smoke and heat from a cooking range (a boxy structure made of fired clay tiles, with holes for cooking pots), or sometimes from a tannur, were drawn down into channels that ran beneath the house floors. The subfloor channels were made of tile, roofed over with slate, then covered with a thick layer of clay. At the far end of the house was an outlet for the smoke. Once the fire died down, the outlet was sealed to retain the residual heat. The clay coating had to be periodically replaced, as it had a tendency to shrink. Changing the coating prevented fumes from seeping into the rooms. Russian archaeologists have excavated a *tawakhaneh* believed to have been in use in 2000 B.C.E.[7]

The Koreans and Chinese did not need bread ovens, as their diets were based on rice, not wheat. However, they constructed similar heating systems in their homes—marvelous raised clay platforms, like great mud-brick beds, on which the whole family could sit or recline or perform household chores in warmth and comfort. The platform was attached to a cooking stove–called a *k'ang*–also made of clay. There were holes in the top of the range for the cooking pots or woks and a fire inside. A family would use whatever fuel they could scavenge; leaves, brush, wood, dried dung. Variations on the k'ang included the *ti-k'ang*, which warmed the floor rather than a raised platform, and the *ton-k'ang*, which warmed the walls. The k'ang was popular during the Han dynasty (which began 206 B.C.E. and lasted until 220 C.E. with a short interregnum from 9 to 25 C.E.) and still in use in the seventeeth century.

The Korean *ondol* more closely resembles the tawakhaneh in that heat and smoke from the cooking range is channeled under the floors in order to warm the room above. Sometimes additional outdoor fireplaces are used to add more heat to the subfloor channels. The channels have earth floors and earth or clay sides. Bricks pillars hold up the slate or granite tops. A layer of clay, often mixed with straw, covers the stone, and then two layers of paper are set down. Modern Koreans sometimes use cement instead of clay for the smooth floor, and no longer use layers of paper. As with the t'ang and tawakhaneh, the challenge is to benefit from the warmth of the smoke without suffering from the fumes.

Though the Romans lived in a relatively mild climate, their subfloor heating systems, *hypercausts*, were superior. They built brick-lined channels for heated air to flow under their floors. The engineering was superb and the flue gases never leaked.

By the middle of the sixteenth century, Europe was enveloped in the Little Ice Age. Autumn yielded to the cold of winter sooner than in the past and spring came late. Average temperatures were lower than they had been in centuries (and lower than they are today). The North Atlantic was choked with ice. Glaciers grew beyond the mountaintops of the Alps and invaded the inhabited valleys. Worse, after centuries of clearing, many of the vast forests were now depleted. In the chill of the Little Ice Age, firewood was no longer abundant. Kings, inventors, potters and ordinary people began to think about fuel efficiency and ways to improve their hearths and ovens. In order to survive, families needed to extract as much warmth as possible, for as long as possible, from their fires.

Wealthy Germans built *steinofen* or stone ovens. The ovens themselves were built of fired clay, but above the firebox there was a chamber that was filled with stones or sometimes iron balls. The fire was allowed to blaze until the stones turned red hot, often splintering apart. The stove was clamped shut and then no more fuel was added. The stones and the masonry held the heat long after the wood had burned.

Hungarians turned to straw supplemented with hay, manure, and whatever dried weed stalks they could gather for fuel. They called their stoves *bubos kemence—bubos* meaning "oven" and *kemence* "straw." These starkly beautiful beehive ovens, plastered smooth on the exterior, were direct descendants of the beehive ovens built in eastern Europe thousands of years before. They were usually built between two rooms. The firebox opened into the kitchen, and the round closed dome warmed the adjoining chamber.

Peasant families elsewhere in eastern Europe also clung to their enormous beehive ovens, described by foreign travelers as great lumps of clay that dominated their houses and their lives. Ovens were built above the firebox and gradually the stoves were made of fired brick and covered with tiles. There were doors for the ovens and fireboxes and nooks for candles and the samovar. Some stoves had a few horizontal interior channels for the heat.

Early ovens vented directly into the room, and later into clay-covered basketry bells or hoods that were suspended over them. Similar primitive wattle-and-daub bells were used to collect smoke and soot from open hearths. These bells did not actually vent the fumes outside, but they were better than nothing.

We do not know when chimneys were first used. Venetian records show chimneys being toppled in the earthquake of 1347 so we know that they were in fairly common use by then, and there are indications that chimneys were in use even earlier.

The earliest chimneys were, like the bells and hoods, built of wattle-and-daub and occasionally of stone. The early chimneys themselves posed danger. They collected soot and creosote. The clay would crack as it dried out. Sparks could easily escape and ignite the thatched roof or the sticks within the chimney or the creosote itself might catch on fire. In a matter of minutes a whole house would be engulfed in flames.

These simple but hazardous chimneys were in use for centuries, though by the early seventeenth century, the courts began ruling against them and ordered chimneys to be built of brick. Stewart Dick, who with the watercolorist Helen Allingham documented the vanishing cottages of the English countryside, wrote, "The first approach to a chimney is the canopy of wood and plaster which was erected over the great fire in the manorial hall . . . At so late a date as 1538, the central cover carried straight up to the roof seems to have been the usual method, for in that year Leland, writing about his visit to Bolton Castle, expresses wonder that the flues were carried up the walls. 'One thinge I muche noted in the haulle of Bolton, how chimneys were conveyed by tunnels made in the syds of the wauls betwixt the lights in the haulle; and by this means, and by no covers, is the smoke of the harthe wonder strangely conveyed.' "[8]

Dick continued on the fire risks: "So from a very early date, indeed from the time that bricks came into common use in the fifteenth century, you find everywhere corporations making bye-laws to guard against this danger. Thatch is whitewashed, chimneys are built higher, and most important of all, wood is strictly forbidden in their construction. The earlier chimneys had sometimes been made of wickerwork, and even now [his book was published in 1909] there survive in Surrey and Sussex some curious examples of chimneys in wattle and clay . . . but soon brick became the chief and almost only material in use."[9]

Chimney inspectors were instituted in the British settlements in New England, lest one house fire burn a whole village down. There were monthly chimney inspections in Hartford, Connecticut, by 1635, and similar laws were enacted throughout the colonies. Fire, though a necessity of life, was a constant threat to it.

Dick points out that, in England, violations were flagrant and frequent: "In the court-leet of the Borough of Clare, Suffolk, in an entry dated 17th August 1621, ordering, 'John Skinner of Sudbury who has a dangerous chimney to amend it.' Then comes the following: 'Item. We, the chief inhabitants and headboroughs of Clare, at the present court assembled, doe forever hereafter, for the good of the estate of the towne, doe conclude, order, and agree, that no man shall erect, and build up any chimney within the borough, but only of bricke, and to be builded abov the roof of the house fower feete and a halfe.'" The borough went on to issue a fine. But one hundred years later, on April 7, 1719, the same borough issued another ruling: "'Item. We present Mr. Jno Smith, clay chimney, and Mrs. Grace Lugdens, clay chimney, and ye Town House chimney where Jno Martin live, if not all made new brick chimneys, within 3 months do fine them 10s a piece.'"[10]

The interiors of brick chimneys were coated with a lime or clay plaster to seal the mortar and joints against creosote and air leaks. Sometimes the top of the chimney was closed, with side slits for the escaping smoke. And, in some instances, the chimneys themselves became works of art, with fancy brickwork. Often several flues rose together up through the center of the house, creating a stunning effect. Dick described them as a "collection of joined shafts with many angles," and said they gave "a variety of light and shade. The head of each shaft in the finer examples [was] formed of projecting courses of bricks; sometimes . . . specially molded" though "delightful results" could be "obtained by the use of ordinary brick" too. In the American colonies, a single, massive center chimney, sometimes accommodating more than one fireplace, was utilized rather than the British style cluster of multiple center chimneys.

By the Victorian era, as coal replaced the diminishing supply of firewood in England, chimney pots were added to the tops of chimneys. The earliest chimney pots were made in the fourteenth century, usually as a protection from wind and rain, but they were rare. With the advent of the Industrial Revolution, potteries began to manufacture them in vast quantities. Made of fired clay, chimney pots protected the hearth below from rain and snow, added height to the chimney and (theoretically at least) prevented downdrafts, a particular menace of coal fires. Pots were attached to the chimney top with mortar that now and then needed reattaching lest they tumble down in a high wind and shatter on the ground below, or worse, strike a passerby. Chimney pots were always decorative, with scallops and rings, fancy crowns, and sometimes various oddly shaped protuberances. Many of the design innovations were believed to make the pots more effective.

By the middle of the twentieth century, most chimneys were built with factory-made clay flue liners. These liners, made of extruded clay and usually rectangular in shape, were stacked up inside the chimney as it was built. Today, many chimneys are lined with triple-walled metal pipe, or the entire chimney itself is made of stovepipe; but though attractive for its initial lower costs, metal is not as long lasting as brick masonry chimneys and ceramic flue liners.

Chimneys were also added to the closed ovens and stoves of eastern and central Europe, again beginning with simple clay or wattle-and-daub chimneys and progressing to brick. They rose up from the base of the oven, at the back, and never from the top, as the weight would be problematic.

Early chimneys did not always reliably draw smoke out of the fire and the house; the relationship of size and height to draft was little understood. Nevertheless, they worked better than no chimney at all, and as chimneys became more prevalent, rooms became less smoky. However, along with smoke and fumes, precious heat was drawn up the flue. In the milder climate of the British Isles, this was an acceptable compromise, but in the colder regions, it was a problem.

Households with enough means had a black stove and a white stove. The black stove offered more warmth because it did not have a chimney. But the room was smoky and sooty, hence the name black stove. A family gathered around the black stove would be toasty warm but would suffer from the usual smoke-induced runny noses, stinging eyes, and headaches. The improved white stove featured a chimney that pulled the smoke up and out of the house but the room never got as warm as the room with the black stove.

The first stove tiles were vessels thrown on the potter's wheel. When affixed to the stove, gaps remained between their rims. As stoves became more boxlike, potters squared the tiles so that they would fit closer together and better cover the entire stove. They were put onto the stove with their bottoms facing outward. The walls of the tiles created an insulating air space between the tile and the stove.

Almost no self-respecting home decorator or potter, even in those early trying times, could resist decorating the plain square tile bottoms, and so they were glazed and embossed with lovely designs. It is unclear when tiles were first used to cover stoves but the idea probably dates back at least to the fourteenth century and likely earlier. They appeared in Sweden and eastern and central Europe.

One winter day in January 1767, the king of Sweden, frustrated with the ongoing tribulations of fuel inefficiencies and shortages, heat loss, and the poor quality of indoor heated air, turned to Baron Carl Johan Cronstedt and asked him to come up with a solution. Cronstedt enlisted his friend Baron Fabian Wrede to help with the project. By the following fall, the men had drawn up plans for tile or plastered brick stoves with interior vertical channels. These channels would safely keep the smoke and the heat inside the stove for a long time before releasing them up the chimney. The longer the hot gases stayed in the channel, the more heat they imparted to the masonry.

Cronstedt explained, "When building the air channels you must recognize that the thinner they can be made, the better, and the more rapidly they will take and deliver warmth. For this I have a potter make thin bricks, lapped on the uppermost and lower edges, so that all cross-joints can be made strong and tight, which is rather important so that no smoke may slip from the smoke channel, in which case an inconvenient smell would enter the room."[11] Cronstedt and Wrede's design called for a sturdy foundation and included built-in clean-out holes for removing ashes. They thought that a quick hot fire made of uniformly sized finely split logs would be best and recommended that as soon as the fire burned down, the damper should be closed tight. This would help retain heat and save on fuel. Again, Cronstedt explained, "Our woodlands are so reduced that there is fear that fuel will be insufficient. We must abolish heavy wood consumption."[12]

The designers made wooden models of their new stove so that potters could make tiles that exactly fit. By 1775 they offered plans to the public and soon their fuel-efficient designs caught on in neighboring countries.

Tile stoves not only offered clean reliable heat, they became works of art. Most tile stoves were square or octagonal or hexagonal but some imaginative stove builders took advantage of the plastic qualities of clay and sculpted fanciful stoves including an Austrian stove in the form of a plump woman with a basket on her head. Tile makers painted flowers and vines, even topiaries onto their wares and added three-dimensional embellishments. Glazes for the stove tiles were vibrant blues and greens and soft glossy whites.

Not everyone thought that these new tile stoves were pleasing to look at. Mark Twain, writing in 1860, expressed his displeasure. He was touring Germany and complained of "one of those tall, square, stately white porcelain things that looks like a monument, and keeps you thinking of death when you ought to be enjoying your travels."[13] It didn't matter to the American humorist that the stoves were efficient; he thought they were most unattractive.

Two and a half centuries earlier, in 1613, Englishman Fynes Morison was also critical of the tile stove. He wrote, "The intemperatenesse of colde pressinge great part of Germanie, instead of fier they use hot stoves for remedie thereof, which are certaine chambers of roomes having an earthen oven caste into them, which may be heated with a little quantity of wood . . . they keepe the doores and windows closely shut, so as they using not only to receive gentlemen into these stoves, but even permit rammsh clownes to stande by the oven till their wet clothes be dried and themselves sweate; yea to endure their little children to set upon their stooles within this close and hot stove (let the reader pardon my rude speeche as I bore with the bad smell); it must needes bee that these ill smells, never purged by the admitting of any freshe aire, should dull the braine, and almost the spirits of those who frequent the stoves."[14]

The British never embraced closed stoves. They wanted to see the dancing flames and the flickering light that played on the walls. They wanted to hear the crackle of the wood as it burned. They liked their enormous open hearths, built of bricks (and sometimes stone), and liked to draw their chairs close, often inside the fireplace itself. Here they would read and do chores and cook their meals.

I have a wonderful old engraving on my study wall of an early English fireplace. There is no separation between the floor and hearth. They are both made of rough bricks laid without mortar, the hearth flowing into the floor. Next to the fire there is a carved chest and a large earthen cooking pot. An old woman, a kerchief knotted at the back of her neck, a shawl drawn across her shoulders, and an ankle-length apron tied over her long-sleeved dress, sits very close to the fire on a low wooden chair. She has a heavy book on her lap, and she is reading to a little girl who sits in rapt attention directly in front of, almost inside, the fireplace. I do not know who the artist is. I found the picture in a local antique shop and bought it for a few dollars. But I love looking at the old woman and the little girl (her grandchild?), clearly poor, but happy with a book and an open fire. The picture epitomizes, I think, the relationship the English had with their hearths, a relationship that went beyond the necessity of a place to cook their meals and warm themselves. They brought this affection for the open hearth with them to the New World, where the climate was harsher, but fuel was abundant.

■

IN NEW ENGLAND today, real estate ads for antique houses always include "original beehive oven" in the advertising copy if indeed the oven has not been boarded up or torn out during a "modernization." It's likely that very few twenty-first-century house buyers actually plan to bake their daily bread in an eighteenth-century brick oven, but if they like the idea of buying an old house, they usually like the idea of a house with a beehive oven. The massive brick fireplace, the cranes for pots, and the arched oven on the side give an old cape or saltbox the charm and romance of days gone by and, as the agent knows, adds dollars to the selling price. Bread is "the staff of life"—we earn our "bread and butter" we "break bread" with our friends and family. It is this long-held association with family and nourishment that lends antique brick ovens much of their appeal.

Today, many pizza houses, both chains and locally owned establishments, boast brick ovens, usually fired with gas, sometimes continuously, to meet heavy demand. The kitchens of television cooking shows are often equipped with brick ovens and even open hearths for grilling. Viewers can catch a glimpse of the ash and coals of the fire as the celebrity chef removes a skewer of caramelized onions or "artisan" bread. The paper packaging of breads for sale in grocery stores often claims that the loaf is hearth-baked or, in one popular brand, the square, uniformly sized loaves are labeled "brick oven" bread.

For the colonial housewife, however, baking bread in a brick oven had nothing to do with choice. It was the only way.

When they first arrived on the shores of New England, wood was, in the settlers' minds, almost too abundant. The dense forests of oak, pine, maples, many of the trees with stout, very tall trunks, had to be cleared for fields and pastures, arduous work. The cut wood was used for building houses and barns, for fences, and for fuel.

During the long cold winters, colonists burned enormous quantities of firewood in order to keep their thin walled houses warm. "A typical New England household," William Cronon writes in *Changes in the Land*, "probably consumed as much as thirty or forty cords of firewood per year, which can best be visualized as a stack of wood four feet wide, four feet high, and three hundred feet long; obtaining such a woodpile meant cutting more than an acre of forest each year ... It is probable that New England consumed more than 260 million cords of firewood between 1630 and 1800."[15]

Inevitably, the seemingly never-ending abundance of firewood became a shortage as New England was transformed from a landscape of forest to one of pastures and fields. Now colonists, like their European counterparts, became interested in conserving fuel. Coppicing never caught on in New England, and there weren't other alternative fuels such as peat. Ironically, it was an unpopular British sympathizer who found a solution—ironic, because the British, though suffering from wood shortages themselves, clung to the romance of their wasteful open hearths, and because the man's loyalty was to the crown.

While George Washington was battling the British, Benjamin Thompson was packing his bags. Twice his New England neighbors had accused him of secretly siding with the Tories, or worse, spying.

He feared, not unreasonably, that they would kill him.

Thompson was born in his grandfather's farmhouse in Woburn, Massachusetts, in 1753. When he was two years old his father died tragically, leaving his son to grow up in hardship. Despite this, and despite his lack of formal schooling, Thompson demonstrated a quick and inquisitive mind at an early age. He was curious and liked to figure out how things worked; he liked to take things apart and put them back together.

When only nineteen, Thompson was invited to teach school in Bradford, Massachusetts. Within a few months, the local pastor, Reverend Timothy Wilton, also formerly of Woburn, was so impressed with the youth's intellect that he asked him to move to Concord, New Hampshire, to teach and continue his scientific studies.

The young Thompson had been in Concord only a short while when he began to court his mentor's daughter, a wealthy widow fourteen years his senior. They soon (too soon, according to some folks in town) married. Whether Thompson was lonely for companionship so far from home, deeply in love with his more mature bride, or, as the gossips speculated, attracted to her land holdings (she owned two-thirds of the town, inherited from her deceased husband) we can only speculate, but he openly, one might say conspicuously, enjoyed his newly acquired wealth. He quit his teaching position and made a great display of living the aristocratic life. This annoyed the other residents of town. A few local patriots accused him of supporting the British, which, with his comfortable life, historians believe he probably did. When Thompson suspected that his neighbors planned to take action against him, he slipped away from his wife and commodious home, stole his brother-in-law's horse, and, under cover of night, raced back home to Massachusetts undetected.

But in Massachusetts, too, he was accused of being a Tory and, more serious, a spy. He fled to England aboard a ship of British evacuees. The Crown embraced him, and after a brief sojourn in London, he returned to America as a lieutenant colonel in the king's American Dragoons. His assignment was to form a regiment of horsemen in the king's service, but by then, the Revolution was nearly won (or from the Crown's viewpoint, lost). Thompson had little to do in service of the king and was most unwelcome in the newly formed nation.

Returning to London, Thompson began his long career in scientific inquiry and social and agricultural reform. He traveled on the Continent, where he encouraged Bavarians to grow turnips and potatoes, and founded schools and colleges. He created a six hundred-acre garden in Munich. The elector of Bavaria rewarded Thompson for his good works and service by making him a count in the Holy Roman Empire. He took as his name Rumford, which was the earlier name of the town of Concord, New Hampshire. He must have chuckled to himself at his private joke.

Count Rumford moved back to London and continued his scientific work. He focused on heating and cooking. Rumford understood that most of the heat generated by a wood fire in a typical vast fireplace vanished up the chimney, leaving the room chilly. In both England, where the climate was milder than elsewhere in Europe, and in America, fireplaces were very wide, with flat backs parallel to the front hearth, and straight sides. Ovens were often placed in the interior of the fireplace so that the cook had to step inside to bake her bread or beans. Eventually, ovens were moved to the side, making work a little more comfortable for the housewife.

Nevertheless, though huge quantities of wood were burned, especially in America, there was no thought to warming a whole room with a fireplace. Children and parents might gather close to the flames to warm themselves, one side at a time, or actually sit inside the fireplace, but the rooms themselves remained chilly.

Benjamin Franklin complained, that the "cold Air so nips" one's backside that the room was not at all comfortable.[16]

And despite the addition of chimneys, smoke still often plagued a household. "Cosy as the ingle-nooks were," Stewart Dick wrote in his turn-of-the-century account of old-fashioned English cottages, "they must often have been very uncomfortable to sit in, as the huge fireplace with the capacious chimney was apt to smoke badly. Nowadays [1909], one finds a little curtain hanging from the huge oak beam that spans the fireplace, which to some extent abates the nuisance. Another device was to let the wood ash accumulate on the open hearth until the fire itself was burning a foot or two above the floor-level."[17]

Count Rumford set himself the task of improving the traditional fireplace. He understood that it was the radiant heat, reflected back into the room from the bricks, that created a room's warmth, not the air heated in the fireplace. To maximize this, he explained in an early essay, "the best form for the vertical sides of a fireplace . . . is that of an upright plane, making an angle with the plane of the back of the fireplace of about 1350."[18] He designed a shallow fireplace with the back wall shorter than the front opening and with slanted sides. The fireplace was taller than it was wide. "The best materials," he continued, "...are fire-stone [sandstone] and common bricks and mortar."

To enhance heat reflection, he recommended white washing the bricks and railed against the use of iron firebacks, which had become popular. Iron, as he knew, conducts heat; it does not radiate heat.

This new design did indeed spread warmth beyond the hearth, but the smoke in a tall shallow fireplace has a tendency to swirl out into the room, choking the inhabitants even more than the large fireplaces Rumford was trying to replace with his improvements. He solved this problem by designing a rounded chimney breast. This encouraged the smoke to go straight up the chimney. Eventually, he also reduced the interior dimensions of the chimney.

The Rumford fireplace gave off more heat than other fireplaces, burned less wood, and prevented smoke from entering the house. It immediately became popular on both sides of the Atlantic and by the mid nineteenth century, "Rumford" became synonymous with "fireplace."

Thomas Jefferson installed Rumford fireplaces at Monticello. He also installed another of the count's inventions, the kitchen range. Rumford designed a massive brick range with a cylindrical oven, and holes in the range top into which cooking pots could be inserted. There were multiple fires under the pots. If it wasn't time to prepare a meal, the opening was covered with an earthenware lid so that the fire smoldered slowly, ready to be stoked when it was needed. In addition to the massive brick range and ovens, a Rumford kitchen required only a small (also brick) Rumford fireplace for roasting meats.

The Rumford kitchen was fuel-efficient and cooler and safer for the cook to work in than a large open hearth. Rumford had great success with it in Munich and, using it to cook his special soup, was able to feed the poor for a halfpenny a day.[19] French chefs quickly embraced the range, which some food historians credit with revolutionizing restaurant fare. Now foods could be cooked quickly, just prior to being served, and innovative chefs could offer a variety of dishes on their menus. A smaller version of the range made it possible for a household cook to prepare fancy meals in her own home. French families equipped with a Rumford kitchen could enjoy a break from the usual soups, stews, and roasts and dine on a light omelet or an herbed sauté.

A woman checks a pot on her brick stove.

The Rumford kitchen, however, enjoyed less popularity in the United States, despite the success of the Rumford fireplace. Only wealthy Americans in towns such as Boston and Newburyport installed them in their kitchens. Count Rumford continued to experiment and write about his fireplace. In 1796, he mentioned the experiments that he was conducting and the craftspeople he employed to help him, including, "Mrs. Hempel, at her Pottery at Chelsea."[20]

Today, Rumford fireplaces are again popular. Recent tests confirm their efficiency. And several modern companies, such as Superior

Clay Company, now prefabricate Rumford throats in their large ceramic factories.

Count Rumford's range was not quite as innovative as everyone thought. As we have seen, his mid-eighteenth-century innovations followed thousands of years in which the extraordinary heat retention quality of clay was utilized in cooking and heating. Remarkable similarities in the design of ovens occurred across continents and millennia. There were also interesting design departures. Improvements were forgotten and rediscovered. Still, the principle was always the same: clay retains heat.

Ranges surprisingly similar to Count Rumford's were in use in pre–Han dynasty China, the Sumerian city of Ur, ancient Rome, and Egypt. The Chinese ranges, dating from about two thousand years ago, were made of brick or adobe. They had holes for pots, or clay round-bottomed woks, similar to Rumford's pot holes. The pots fitted tightly into the openings, thus conserving fuel. There was a stoke hole and a separate firing chamber. The range could be up to two meters square. These ancient ranges used little fuel vis-à-vis the amount of food that could be prepared.

The great British archaeologist Sir Leonard Woolley (1880–1960) discovered a "restaurant" kitchen dominated by a massive, solid brick range in Ur (now southern Iraq), dating from nearly four thousand years ago. The top of the range was outfitted with troughs for charcoal braziers. Near the stove was a "circular bread-oven."[21]

Both the Chinese and the Mesopotamian ranges raised the cooking surface to a comfortable working height for the cook, and both were made of fireproof clay. The Chinese stove, however, like Rumford's range, contained the fire within its clay mass, while the stove in Ur held the fire aloft in its troughs.

Romans too outfitted their kitchens with ranges, which, like those in Ur, were essentially raised hearths, fueled by wood or charcoal. And they had sophisticated ovens, as attested to by the writer, H. E. Jacob when he visited the Casa di Salustio bakery in Pompeii in 1923 and wrote, "the oven, the fine modern oven that the people who dwelt in this place two thousand years ago had cleverly constructed … with what care the Roman masters had surrounded the inside of the arched oven with a square hollow space to insulate the hot air. There was a draft for the smoke, a container for ashes, and a container for the water with which the crust of the half-baked bread could be moistened, in order to impart a fine glaze. Adjoining the oven were two rooms for the bakers ... and on the wall a picture of the oven goddess Fornax."[22]

Despite the efficiencies of the Rumford fireplace, iron stoves eventually replaced the open fire in most American homes. Iron offered quick, very hot heat. But these stoves required particular care. Ashes had to be cleaned out regularly. Soot had to be scraped from the pipe as well as from inside the stove. And the stove had to be "blacked." Eliza Leslie, who gave domestic advice in *The House Book: Or, a Manual of Domestic Economy*, published in 1841, explained, "Take half a pound of black lead finely powdered and ... mix with it the whites of three eggs well beaten; then dilute it with sour beer or porter ... Having stirred it well, set it over hot coals, and let it simmer twenty minutes. When cold, pour it into a stone jug, cork it tightly and keep it for use."[23] The mixture was applied with a "soft brush, and then polished off quickly with a clean hard brush." After that, a varnish had to be applied. All this was done when the stove was cold and had been thoroughly cleaned. This was nasty work.

Soon, wealthier housewives were offered stoves enameled in porcelain. Though the porcelain could chip if a rambunctious family member inadvertently banged a heavy iron pan into it, it was very durable and was much easier to clean and care for than plain black iron. The finish, in white, cream, or almond; was smooth and glossy. A stove finished in porcelain enamel and trimmed with nickel dressed up the dreariest late-nineteenth-century kitchen and is now highly collectable.

Today, porcelain enamel is used on most modern steel electric and gas stoves (and other kitchen appliances). It is easy to clean, withstands sustained high temperatures, and "has no pores or scratches to provide bacteria with a place to grow."[24] It comes in a multitude of colors, though white remains the most popular. Though these gleaming modern stoves, fueled by electricity or gas, are distant relatives of simple clay ovens and brick fireplaces, they nevertheless, boast clay—in the porcelain enamel—as a major component. Some electric stoves also have raised flat French burners, which use ceramic insulation. Others have special ceramic glass tops.

In the industrialized world the cookstove or oven is no longer also the source of heat. With the rise of central heating, homes were equipped first with coal hot air furnaces or boilers followed by gas or oil hot air furnaces or boilers. In the later twentieth century, some houses were outfitted with electric heat. But we can't get away from the role of clay: the fireboxes of even the most modern furnaces are lined with firebrick.

THE FIRST MACHINE AND THE
DEVELOPMENT OF AN INDUSTRY

*. . . pot-making is perhaps the earliest conscious utilization by man
of a chemical change.*

–V. GORDON CHILDE

IN THE PROCESS of learning to work with clay, primitive ceramicists made two remarkable technological advances: the potter's
wheel and the kiln. In these inventions, they were finding ways to
make their daily lives better, their work easier. From our place in time,
however, we marvel that they used the principles of thermodynam-
ics, chemistry, physics, and mechanical engineering long before these
disciplines were formally invented. Primitive? Perhaps not.

The act of firing clay in a kiln is the first act that we know of that
purposely effected a chemical change. When we purchase a handmade
set of mugs, we feel we are enhancing our lives with something more
natural than plastic. Incredibly, though handmade mugs *are* made
using a very ancient process, fired clay was in fact the first synthetic
produced!

Why? Because intense heat irreversibly changes clay.

Partially dry clay (called "leather hard," this is clay that can be carved but not bent) is between 8 and 15 percent water. Bone dry clay (clay that can no longer be carved or manipulated) is about 3 percent water. If you drop either of these states of clay into a bucket of water, they will become soft again. However, if you put clay into a fire, once temperatures have reached between $350°F$ and $400°F$ ($175°C$ and $205°C$) all the atmospheric moisture is driven from it. At $950°F$ to $1,300°F$ ($510°C$ to $705°C$) the chemically bonded water is also driven from the clay. At these temperatures, gases from any organic materials and fillers are also driven out. Once the clay is absent its chemically bonded water, it can no longer be returned to a soft state. The clay has been chemically changed by firing.

Different clays "mature"—reach their maximum strength—at different temperatures. Earthenware generally matures around $2,000°F$ ($1,090°C$). Stoneware matures around $2,350°F$ ($1,290°C$). Fired beyond these temperatures, the ware bloats or slumps and loses its shape. The highest temperature at which a clay body retains its shape is its vitrification point. Generally, ware is fired somewhat below the vitrification point. Earthenware remains porous but stoneware and porcelain become nearly nonporous.

Other chemical changes take place when clay is exposed to intense heat.

At temperatures as low as $1,832°F$ ($1,000°C$) crystals of alumina-silica ($3Al_2O_3 2SiO_2$), known as mullite crystals, can begin to develop, giving the clay strength. As higher temperatures are reached, and more mullite is formed, the clay becomes more durable. Before firing, clay consists of minerals such as kaolin or ball clay, plus feldspar, quartz, mica, and various impurities. After firing, in addition to mullite, the body is now several forms of silica, cristobalite, silica-free quartz, and noncrystalline silica. When the fire temperatures reach $1,063°F$ ($575°C$), the quartz crystals change from alpha crystals to beta crystals. They expand. When the fire cools down, the crystals shrink. This is known as quartz inversion and can cause dunting (cracking).

As we have seen, our ancestors probably first discovered that fire could transform clay when they noticed that their clay-lined hearths turned to hardened basins after use or when the little animals and people that they had modeled from sticky river mud, turned to stone when left in the hearth. There is, however, a vast difference between a cooking fire and a fire suitable for ceramics—the family meal would become a mere residue of ash if prepared in a fire hot enough to produce ceramics!

Ceramics require a fire that begins gently and then rises to an inferno of red heat, at minimum, and yellow or close to blazing white heat at optimum. If you look inside the fire as it reaches temperature—and potters have done just this for thousands of years—you will see the pots or sculptures luminously glowing first red, then bright yellow in the swirling gaseous flames. It is a blinding, awe-inspiring sight, and it is fraught with dangers for the ware—and sometimes for the person attending the fire.

If there is moisture in the object to be fired, the piece will explode.

If the temperature climbs too fast, the items can crack or break.

If cool air rushes into the still warm fire, the pots can crack or break.

The fire must be cooled slowly at least until the temperatures descend to 400°F (205°C), otherwise the ware will shatter. If the temperature changes too quickly at the moment of the change from alpha to beta quartz crystals, or back, the ware will crack. There are hundreds of things that can go awry in a firing; the fire can get too hot and pots can melt, the kiln can collapse, the house can catch fire!

No wonder the sites of pottery kilns, throughout history, throughout the world, are strewn with "wasters." These are treasures for the archaeologist, but loss and heartbreak for the maker. Fires, even to a community of clay workers that has conducted hundreds of them, are filled with risk. They are also, the first time and the hundredth time, exciting. They are a time of anticipation, dreams, focus, and hard, sweaty work. Every now and then, the kiln gods reward the potter with a gift of unexpected and unsurpassed beauty, a piece that is better than anything anyone imagined.

■

FIRING IS, AT its core, problem solving. You want to expend as little fuel and energy as possible. You want to prevent the ware from breaking in the fire. And you want to produce something that is as durable as it can be. You want all, or as much as possible, of the ware in the fire to come out nicely. It is a waste of time and energy if half the items are underfired or overfired. This was as true for clay workers in the furthest recesses of history as it is for the engineers in lab coats designing a way to fire tiles for the space program. The construction of kilns is an attempt to control the fire and thus the chemical process that takes place within the flames.

Just how ancient is this knowledge?

From 1924 to 1938 the archaeologist Karel Absolon excavated the site of a late Ice Age community of mammoth hunters nestled near a stream in south central Czechoslovakia. During the dig, he found a startling artifact that profoundly changed our understanding of ceramic history: a beautiful stylized figure of a woman sculpted in clay and *fired*. The woman's exaggerated breasts hang down to her hips, which swell out to almost grotesque proportions. Her legs are pressed close together, tapering to a near point. Dubbed the Venus of Dolni Vestonice, this extraordinary Upper Paleolithic work of art is remarkable because it was deliberately fired almost thirty thousand years ago, fifteen thousand years earlier than anyone had imagined that ceramics were made!

The late Ice Age community where the Venus was discovered consisted of five huts, as evidenced by postholes. The largest hut held five hearths. Around the hearth area, there were ash pits and more than thirty five thousand pieces of flint, suggesting that the hut was a living area and a busy workshop. Further excavations took place in the late forties and early fifties. In 1951, the archaeological team was startled to find a kiln. It was located in a hut that was set apart from the other huts, and built partially into a hill. The hearth in this hut differed dramatically from the stone-rimmed floor hearths in the big hut and was unmistakably used to fire the clay figures. The kiln, open at the front, had thick clay walls and evidence of a domed clay roof. The firing space would have been fifteen inches high. Figurines and bits of broken fired clay arms, legs, animal and human heads, and thousands of fired clay pellets were scattered nearby.

Modeled and fired tens of thousands of years ago, this little clay goddess or fertility figure, known as the Clay Venus Dolni Vestonice, speaks to us across the millennia.

Not only did these Old Stone Age people understand that fire has a transformative consequence upon clay, they actually built a kiln to better control the fire and its effectiveness upon their work. They knew to temper their smooth clay with pulverized bone, to give it more thermal resistance. They knew to build walls to contain the fire and keep out sudden drafts. The site was inhabited for a few thousand years. We do not know if habitation was seasonal or year round. Though the village was sheltered from the fiercest climate of the time, the average temperature never rose above freezing. Life was difficult. Scientists speculate that the artist who modeled the damp clay into wolves and bears and wide-hipped women was a shaman, and that the Venus was a ritual object used in a fertility rite or in a ceremony of goddess worship. There is no way to know for sure. We *do* know that whoever made the figures, whoever the artist was who sat in a bermed, hillside hut and made and fired the Venus of

Dolni Vestonice around 27,000 B.C.E., she knew nothing of mullite crystals or quartz inversions. She did not know that she was facilitating a chemical change in her little goddess. But she clearly understood, empirically, the basic principles of firing clay. She was also among the very first ceramicists—if not *the* first—the world had ever seen. It would still be millennia yet before anyone made a pot!

■

A KILN IS not necessary for firing clay; a bonfire will do. You can spread a layer of grass and brush on the ground, place your pots or figurines on this, and then cover it all with more brush, or if you want a higher temperature, pieces of wood. Once lit, the pile will burn furiously for a half hour or so, and then die down.

Indeed, potters of the American Southwest produce truly exquisite pots that are fired with carefully placed dung cakes. Collectors pay high prices for the works of Lucy Lewis and Maria Martinez and their descendants.

Although you can fire clay in bonfires, they are particularly risky and require great skill to minimize losses. The fire can heat up too quickly, thus shattering the pots. Openings can occur in the layer of fuel, and if these are not quickly closed over with more fuel, they will draw cool air across the red hot pots and crack them. A heavy ember of wood can fall and smash the neck of a jar. A strong wind can suddenly blow in across the field, fanning one side of the fire so that the flames race ahead of the opposite side, thus burning unevenly, or heating or cooling so quickly that the ware breaks. A clear day can unexpectedly turn stormy and rain can beat down on the fire, quenching it or cooling the ware too quickly or reducing the fire to a smolder, not hot enough to mature the ware.

It is no wonder that potters throughout time and in different parts of the world have developed superstitions and rituals to ensure a successful firing!

In Korea today, owners of kilns still practice "the ancient shamanistic ritual traditions of offering a blessing to the spirit of the land and fire before placing a torch to the kiln. An offering of rice cake, uncooked rice, water and wine with a pig or cow's head is prepared on the altar. Candles are placed in the bowls of uncooked rice. A processional dance ceremony with flute, cymbal and *changgo* drum is conducted," Edward Adams writes in *Korea's Pottery Heritage*.[1] He goes on to say that women are often not allowed to participate, which has been true in various cultures, especially if the woman is menstruating. The famous sixteenth-century Italian potter Cipriano Piccolpasso, who wrote the first "how-to" book for potters, suggested "invoking the name of God, take a handful of straw, and with the sign of the cross, light the fire."[2]

It is interesting that potters in far-flung corners of the globe, who have had no contact with one another, working in different epochs, resourcefully reused the pots broken in one bonfire to improve the results of successive firings. Potshards carefully placed around and over unfired pots protect them from the direct lick of the flame and sudden temperature changes.

Jane Perryman describes a modern-day bonfire in Himachal Pradesh, the northern mountain state of India, in her book *Traditional Pottery of India*. An eight-foot circle is made and covered with a layer of rice husks followed by a three-inch layer of pine needles. The unfired pots are then set down, mouths facing inward. Ten basketfuls of dried buffalo dung are then placed over and between the pots. There are "several layers of the large water pots, filling in the gaps by inverting smaller and differently shaped pots and continuing to place dung in and around them until the pile is waste high. The pile is covered with more dung cakes ... [and] pine needles up against the side. Sherds are placed around the circumference and then built up to cover the pile, whole potshards around the bottom and curved broken pieces higher up where the pots are smaller." The potter then "covers the outside with rice straw to a thickness of about 20 cm (8 inches) ... and compresses by beating down with a bamboo pole ... sprinkles water on the ground around the circumference to prevent the fire spreading outwards ... the rest of the straw over the pile ... covering the pile with ash to prevent the straw from flying about."[3]

Similar, very ancient firing practices, but with indigenous fuels, are found in Nigeria, Africa, New Zealand, the southwestern United States, and elsewhere in India.

Equally as effective as placing potshards over, and sometimes around, the unfired ware is the practice of digging a hole for the bonfire, or, more properly, pit fire. A shallow pit protects the fire somewhat from gusts of wind and, with the fire in the bottom of the pit, the heat can rise up through the ware. With a deep pit, the fuel and pots can be layered, and in many cases, the whole fire pit covered over with broken pots, and possibly, at the end of the firing, with earth or clay.

Open pit fires and bonfires are fairly fuel-efficient, require little capital investment in the way of building, and, if you understand your climate, your clay, and your fuels, can be very effective, though they rarely achieve temperatures beyond 1,300°F (705°C) to 1,650°F (900°C). They have remained in use for thousands of years, and, in many parts of the world, have continued in use even after more permanent kilns were invented.

But higher temperatures produce stronger ware. Much thought and effort has been expended over the years to create hotter and hotter fires. Temperatures can be reliably increased perhaps 180°F (80°C) by digging air holes to the pit. By adding clay, brick, or rubble walls, and raising, and keeping, the ware above the fire, even higher temperatures can be reached than in the simple pit kiln. In his classic *Kilns: Design, Construction and Operation*, David Rhodes explains that this early "form of the kiln was essentially a cylinder, open at the top, with an entrance tunnel for the fire provided at the bottom ... The floor or platform on which the ware was set was perforated with holes to let the fire pass upward. In some cases this floor was built of large fired clay bars wedged across the cylinder, with clay partially filling in the space between the bars. The ware to be fired was piled into the cylindrical chamber. The fire was built in the entrance and the flame and hot gases from the fire passed upward through the ware, escaping from the top. The top of the kiln could be partially closed off by a loose thatching of broken pottery or tiles. This design represents a great advance over the pit fire, incorporating all the elements of the kiln as we know it today."[4]

Kilns such as these, in which the flames and heat rise straight up, are called "updraft" or "vertical" kilns. We know from Egyptian temple art

One of the most ancient kiln designs is the updraft kiln. This one, being stacked by a potter in India, will be covered with shards once it is full, and then a fire will be lit in the bottom.

that simple updraft or vertical kilns were used by the middle of the third millennium B.C.E. These were loaded from the top, and domed over for each fire. Archaeologists excavated the fourth-millennium B.C.E. remains of a vertical kiln in the ancient city of Susa (in modern western Iran) that was six feet across, considerably more substantial than the Egyptian kilns. Several lovely Corinthian wall plaques have been found that illustrate ancient Greek updraft kilns with permanent domes and side doors for loading. These had protruding fireboxes. In one plaque, the naked potter stokes the fire while ash and smoke belch from the flue opening at the top of the dome. In another, we can see the vases tumble-stacked inside the kiln.

Simple updraft or vertical kilns were prevalent in the ancient Near and Middle East, around much of the Mediterranean, in

Europe, and in the Congo and they are still used in India, Cyprus, Morocco, Spain, and elsewhere. Perryman describes a contemporary kiln of this type in use in India, which is twelve feet (3.7 meters) in diameter and four and a half feet (1.4 meters) high. It is "built from a single thickness of fired brick covered with cow dung mixture. It rests on a platform 2.5 feet above ground level, the platform containing a stoking hole, which leads into a hollow underneath the floor of the kiln. The floor is slightly dome-shaped, with gaps left in between the bricks to enable the heat to rise up."[5] After the kiln is stacked with pots, it is capped with potshards and a piece of sheet metal and then fired with sawdust.

In the New World, ware was most often fired in bonfires but kilns were also invented and used as early as three thousand years ago. The Illinois archaeologist Izumi Shimada discovered clusters of one- and two-chambered keyhole- and pear-shaped pit kilns dating back as early as 1260–960 B.C.E. "Kilns were oriented north-south to northwest-southeast (with the principal chamber on the northern end), matching the prevalent diurnal wind directions of the region."[6] Shimada attempted to replicate the ancient firing process and found that he could reliably reach temperatures of 1,470° F (800° C). Shimada also infers from the evidence that the pits were domed over.

It appears that even in areas where kilns were built, bonfires continued to be used in rural areas and for household use. Kilns allowed commercial activity and production on a larger scale than a bonfire, with better and more reliable results. But they necessitated a serious financial investment and technical skills. Producing a few pots for your family's needs, and perhaps a few neighbors', would not justify the labor and expense of a permanent kiln. In the West, updraft kilns remained essentially unchanged until the Industrial Revolution.

Though the principles and challenges of firing were identical to those faced in the rest of the world, in Asia ceramics followed a remarkable and very different path. By 1000 B.C.E., Chinese kilns were exceeding 2,000° F (1,100° C), higher temperatures than were achieved elsewhere for thousands of years. Some experts believe that the very earliest kilns in China were walled vertical kilns. A Neolithic kiln discovered in Banpo (Pan-p'o) consisted of a fire pit dug into the lower portion of a hill. Flues connected it to a firing chamber above, also within the hill. This was a vertical, updraft kiln dug into the earth.

But the extraordinary contribution of the ancient Chinese was their revolutionary invention of horizontal kilns capable of reaching very high temperatures. The first horizontal kilns were made by tunneling into a hillside to create an inclined interior chamber with a flue opening out the top and a fire mouth at the base. The earth was relatively free of stones and ledge, so the Chinese did not encounter the problem of heat-damaged rocks splintering and crashing onto the pots, as would have happened in a stony or gravelly location. The soil, rich in refractory clay, hardened into a nice ceramic shell. The thick layer of earth surrounding the chamber provided highly effective insulation.

It was the horizontal design and the incline that offered the most benefit. The incline created a strong draw that pulled the flames through the kiln in a powerful cross draft. Relatively even, very hot fires were achieved. By the late Neolithic period (late fifth to early fourth millennium B.C.E.) these kilns were reaching temperatures approaching 2,000°F (1,100°C).

By the beginning of the Common Era—probably before—Chinese potters were building similarly shaped kilns, still on an incline, but now partially above ground. The firebox was dug out below the front of the ware chamber. The chamber was an exaggerated egg shape, with the front dramatically higher than the rear. The rear opened into a flue to the chimney. In northern China, these kilns had two chimneys at the rear and are called "horseshoe kilns."

Single-chambered "climbing" kiln technology spread to Korea and then to Japan, where it was used to produce wonderful high-fired ware for hundreds of years. The climbing kiln of the Silla period (57 B.C.E.–936 C.E.) in Korea was "made by excavating a shallow ditch, 30–40 meters in length, maybe two to three meters wide on a thirty degree slope. The sides are built up with brick and clay to form a tunnel over the ditch. If the terrain was flat the slope would be constructed artificially ... Kiln firing often took a week [and] massive quantities of wood were used."[7] By the fourth century of the Common Era, word of this new and improved Korean kiln technology had spread to southern Japan. The ruling family of Yamato sponsored an expedition to investigate and brought back knowledge of the *ana-gama* (or "cellar kiln"), which was quickly adopted by the Sue potters and used for centuries.

By the Koryo period of Korea (918–1392 C.E.), the tunnel or chimney kiln evolved into "split bamboo" kilns, so called because they resembled bamboo stalks sunk partially into the earth. The interior of the split bamboo kiln was divided into chambers and was easier to control and fire than the single-chambered climbing kiln. This innovation also spread to Japan where it was widely adopted.

By 1000 C.E. potters in southeastern China had begun to build their remarkable "dragon" kilns. Using local refractory clay bricks, they constructed connected, domed chambers on an incline, with each chamber a step above the previous one. The number of chambers varied from kiln to kiln, but it was not unusual for eight to twelve chambers to be built up a hillside. Kilns of this enormity could hold twenty to twenty-five thousand pieces.[8] The pieces were placed inside saggars, cylindrical fireclay boxes, which were then stacked to form great columns inside the kiln. Saggars protected the ware from falling ash or timbers and made loading the kiln easier than piling the pots inside.

The draft created in climbing or dragon kilns was so powerful that chimneys were not needed. The lowest chamber was fired first.

Throughout history, the peoples of China, Korea, and Japan exchanged their advancing skills and knowledge in the production of ceramics through trade, piracy, expeditions, migrations, wars, and invasions. Ideas generally flowed from China through Korea to Japan. In 1592, the Japanese shogun Hideyoshi actually captured Korean potters themselves and brought them back to the southernmost island of Japan to work. This came to be known as the "Potters Wars." The POW potters worked for the *daimyos* (regional lords) producing wares for their households, and for the tea ceremony, which was all the rage, as well as pottery for residents of the surrounding area. Often, one or more of the middle chambers of the climbing multichambered kiln (*nobori-gama*), where the kiln's best results were produced, were reserved for the exclusive use of the daimyo.

The innovations and principles of the horizontal downdraft kilns of the Orient remain unsurpassed today, though there have been improvements in efficiency and ease of operation, particularly in the latter half of the twentieth century.

■

EUROPEAN POTTERS FOLLOWED a different path in their kiln development.

Around the Mediterranean, the simple updraft kiln remained in use. In England, however, the cylindrical updraft kiln grew to a two-story high, aesthetically pleasing bottle shape. Sometimes the kiln held two chambers, one on top of the other. A brick hovel, or second bottle, encased the kiln. This served as the chimney and work area. The hovel was easier to build and less costly than had the weight of a chimney been added to the kiln itself.

The Germans built enormous round kilns with domed tops and a tall chimney off to the side. Fireboxes ringed the kiln. Interior bag (baffle) walls pushed the flame up. The height of the chimney and ensuing draft pulled the flame down, through the kiln interior, to the flue channels, before pulling them up and out of the kiln. Variations of this downdraft kiln fired evenly and well.

During the Industrial Revolution, the Germans and English also built long, rectangular downdraft kilns with arched roofs. Named for their towns of origin, they were respectively, the Cassel kiln and Newcastle kiln. They were used to fire brick and inspired the partially subterranean "groundhog" kilns of the American South.

The late kiln authority Daniel Rhodes wrote that the downdraft kiln "may be considered the ultimate development in fuel burning kilns."[9] The downdraft kilns of the Orient produced beautiful high-fired pottery equal to, and in the eyes of many connoisseurs, surpassing, anything made today. It was more than seven hundred years before Europeans discovered the superior firing qualities of the downdraft design and produced ware of as high a caliber as Asian pottery. Improvements continued to be made, but they were not as significant. In the mid eighteenth century, the French began experimenting with a "tunnel" kiln. With this concept, the fire would always burn. Cars, stacked with ware, rolled on tracks (or in one somewhat harebrained early-twentieth-century version, on barges!) through the kiln. During the passage, the car moved through preheat, heat, and cooling stages. Such a design presented many challenges and it was a hundred years before, in 1877, a Mr. P. Bock built and patented a successful tunnel kiln. Tunnel kilns, rather than intermittent kilns, are widely used in the ceramics industry today.

Until the nineteenth century, kilns were fired with wood. Then coal, particularly in England, was used. The following century saw the introduction of gas and oil and, after World War I, electricity.

Today, many studio potters use cube-shaped electric kilns constructed of brick encased in a steel jacket. Yet there has been a growing resurgence of interest in wood firing: studio potters in Australia, New Zealand, the United States, England, and elsewhere have built climbing kilns and single-chamber kilns fueled with wood and often make their firings a festive community event with days and nights of stoking. My own kiln is a variation of the popular Brookfield design (yes, named for a town), a downdraft kiln with a sprung arch roof, fired with four propane burners. It has a thirty-seven-cubic-foot chamber and a sixteen-foot chimney. I fire slowly and reach 2,336 ° F in about twenty-four hours.

■

ASCERTAINING THE TEMPERATURE inside the kiln is always critical.

An experienced kiln tender can tell what the temperature is reasonably accurately by carefully peering inside at the interior color—red, orange, or yellow—but there is still some uncertainty until the kiln is cooled and opened and the pots are removed. Draw rings were one early solution to this dilemma: rings of clay were placed into the kiln along with the pots, and pulled out while the firing was still in progress. By looking at the ring, the tender could tell how far along the firing was.

In 1886 a German, Dr. Hermann Seger (1839–93), who worked at the Berlin Royal Porcelain Factory (which was founded by Frederick the Great in 1763) invented the pyrometric cone. By this time thermocouples (metal devices for measuring temperature) were known, but they could not withstand the extraordinary temperatures inside a kiln. Seger's brilliant idea was to make little "bowling pins," later known as cones in English, of varying compounds that melted at different temperatures. Cones (which are actually not cone-shaped at all, but skinny triangular pyramids) measure heat work, and are very proficient for the purposes of ceramics. They are placed inside the kiln on a pad of clay, usually in sets of threes: the warning cone, the end cone, and the guard cone. When a cone bends over so that its tip touches the pad, it is considered down. When the end cone is down, the kiln has reached the desired temperature. Modern industrial kilns are equipped with thermocouples, galvanometers, and electronic devices but Seger's cones (which were later improved upon by Edward Orton) are still important measures of heat work. The realization that fire could chemically transform clay into a stonelike, waterproof product and the invention of kilns to control this transformation represent critical progress for civilization.

So does the invention of the potter's wheel.

■

AUTHORITIES DISAGREE ON where the first potter's wheel was used: China, the Fertile Crescent, or Egypt. It is also unclear whether wheels were used for transportation or for pottery first, but most scientists believe that potter's wheels likely came before or were simultaneous with wheels for carts.

The fast-turning potter's wheel enabled potters to make perfectly symmetrical pots with less personal energy and more speed than a coil- or slab-made pot. Watching a skilled potter throw on the wheel is mesmerizing. The potter throws (hence the term *throwing* for the process of making pots on a wheel) a well-prepared ball of clay onto the wheel head so that it sticks. With the wheel in motion, the potter dribbles water on the spinning lump of clay and places both hands lightly around it. The clay spins between the potter's hands, the centrifugal force pushing outward as the potter's palm or palms push inward. Within minutes, the clay becomes centered, a smooth cone of clay so symmetrical that its motion is almost imperceptible. More water, and now the potter thrusts one or both thumbs into the middle of the spinning cone and pulls outward, forming a thick-walled low cylinder. Keeping the spinning vessel wet, the potter, with one hand inside the cylinder and one outside, pulls the clay upward, and the walls thin and rise, the cylinder growing tall as if by magic. The potter can push out with the interior hand and the pot will swell, or in with the exterior hand and the pot will narrow. A vase that would take hours to build with coils appears within five or ten minutes on the wheel.

It is the centrifugal force that is created, and the fact that the wheel continues to spin on its own while the potter throws, that makes the potter's wheel work. Prior to and concurrent with throwing wheels were turntables. These were rotated slowly as coils were added, a pot was paddled, or decorations were inscribed.

Simple early potter's wheels took several forms. The wheel might have a socket or cup at the center of its underside, into which a pivot was fitted. The pivot would be stuck into the ground or floor, and the wheel head would be set in motion by hand, or by using a stick set into a turning hole, before the potter began to throw. It could be the reverse, with the pivot extending down from the underside of the wheel head, into a socket. In both cases, the wheel was hand powered and would spin freely. Lubricants, such as animal fat, could be used in the socket. Illustrations of the pivoted disk potter's wheel appear in Egyptian tomb art just prior to 2,500 B.C.E.[10] Both of these types of simple wheels continue in use today—particularly in India, Japan, and Southeast Asia—and are used to make perfectly thrown and vibrantly beautiful pots.

A Middle Kingdom (1900 B.C.E.) potter's workshop. One kiln is being fired, another is being loaded or unloaded. One potter works at a wheel (far right) while another does a little handbuilding (far left). Clay is wedged, pots are carried, and finished pots are lined up.

The "double" or "kick wheel" uses pivots and sockets and a bearing. The kick (or flywheel) gives added momentum and thus keeps the wheel turning longer. A stone bearing for such a wheel was discovered during the 1930s in the remains of an ancient Canaanite potter's workshop which had operated in the late Bronze Age in Lachish, southwest of Jerusalem, between 1200 and 1150 B.C.E. The well-equipped workshop appears to have been a commercial endeavor; archaeologists found a store of prepared clay, colorants such as yellow ocher, a variety of tools, and an inventory of forty or so finished pots. The flywheel was set in a square pit in the floor, with the wheel head slightly above floor level, so that the potter could sit and work. This type of wheel continues in use in the Middle East and parts of western India today. By 1500 B.C.E. Egyptian tomb art depicted similar kick wheels, though not set into the floor.

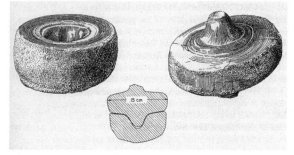

A potter's wheel bearing found in Jericho, now in the Archaeological Institute, University of London, England.

My wheel, built by Lockerbie, is a modern version.

The heavy flywheel is made of cement and the frame and wheel head are steel. It is equipped with a wooden bench for sitting. Two or three good kicks with my right foot, and the wheel has enough momentum to last several minutes, often longer. Even Sam and Sydney, my slim grade-school-age nieces, can easily keep it spinning.

Modern innovations include a treadle wheel, hydropowered wheels, and, popular with studio potters today, electric wheels. But whether powered by an assistant, oneself, or electricity, the principles are the same.

■

ANCIENT POTTER'S WHEELS do not generally survive the centuries, so it is unlikely we will ever discover the first wheels. However, pots and potshards do survive, and by examining them scientists can determine whether they were thrown or made by some other method. Throwing leaves rills or throwing rings in the walls of a pot, which, even if smoothed out afterward by the potter, can be seen when radiographed.

Using the evidence in the pots themselves, V. Gordon Childe gives these dates for the potter's wheel:

Sumer 3250 B.C.E. (± 250 years)

Mediterranean coast of Syria and Palestine 3000 B.C.E.

Egypt 2750 B.C.E.

Harappa and Mohenjo-Daro 2500 B.C.E.

Crete 2000 B.C.E.

Mainland Greece 1800 B.C.

Southern Italy 750 B.C.

Upper Danube and upper Rhine basins 400 B.C.

Southern England 50 B.C.

Scotland A.D. 400

The Americas A.D. 1550.

Writing in the fifties, he considered the archaeological evidence for India and China too scant to know dates for certain. We should note that in the New World, the wheel was unknown for both pottery and transportation until the arrival of Europeans.[11]

Radiographs show that the wheel was used by the Lung-shan in northeastern China during the third millennium B.C.E. They made astonishingly thin-walled, long-stemmed black goblets that would not look out of place at a festive party today, and indeed, archaeologists speculate that they were used for special occasions or ceremonies. They also produced more traditional jars, jugs, cups and bowls.[12]

Unlike bonfires and kiln building, unlike coiling, paddling, and slab building, where the same ideas and solutions appeared independently in different parts of the world, throwing on the wheel was more likely a skill that passed from one group to another, a revolutionary concept that traveled thousands of miles. It may have been invented more than once by more than one culture, but if so, it's not likely that this occurred often. Throwing on a fast wheel is not an obvious thing to attempt and takes considerable practice before mastering. Seven years is the norm. In Japan, an apprentice must throw hundreds of bowls before one is considered good enough to keep!

The potter's wheel is the first machine to use true rotary motion. That is, it could turn in one direction indefinitely. When an early Neolithic man rolled a stick back and forth between his palms, with the end pressed hard upon a board, so that he created a hot coal, he was using partial rotary motion. But until the invention of the wheel—and remember, most scientists believe that the first incarnation of the wheel was the potter's wheel, followed by wheeled carts—continuous rotary motion was unknown. Rotary motion is the principle upon which the great inventions of the Industrial Revolution were based. It led to the miller's waterwheel, the steamship, the airplane and jet, the clock, cogs, the automobile, the motor, and a thousand other inventions that have improved or changed our lives.

With the wheel, potters could greatly increase their output. A different set of skills was required than the skills needed for coiling or paddling a pot. Pot making became a commercial and professional occupation and largely the purview of men. Moira Vincentelli, who has spent much of her career looking at pottery and gender, writes, "It is very easy to assume that household production of ceramics has always been undertaken by women and although there are exceptions, the ethnographic record provides strong supporting evidence for this . . . Historically, what we know now is a world where almost everywhere specialist pottery has emerged at some level. With the exception of large parts of Africa, in every continent men have been producing pottery on the wheel for many centuries and in some cases millennia."[13]

Childe points out that "a precondition for the use of the wheel is a social surplus sufficient to support the potter and his family—that is an effective market for his wares. Pots, being fragile and bulky, cannot normally be exported under primitive conditions of transport over any considerable distances. The market must be local . . . it would not pay to set up a wheel in a village of less than a couple of hundred households."[14] This meant that the potter's sons would have to move to other villages to set up their shops and by moving they would spread knowledge of the wheel to more people.

■

Two PIECES OF pottery equipment, the kiln and the wheel, which today may seem unremarkable—even, in the twenty-first century, quaint—are in fact two of the most important inventions in history, inventions that form the foundation for countless others critical to modern life. And both of them came directly from the use of clay.

Unusual representation of a female potter from a mid-fifteenth century deck of cards presented to the Hapsburg court. Though the potter is incising decorative lines into the pot rather than throwing, the image conveys the energy of the wheel and the exuberance the potter feels. Note her bare feet and the pile of clay.

SET THE TABLE

From a Simple Bowl to a 2,200-Piece Dinner Set

*In the beginning God gave to every people a cup of clay,
and from this cup, they drank their life.*

—NORTHERN PAIUTE PROVERB

As long as humans have been serving food on ceramic dishes, we have been concerned with the appearance of those dishes. Tableware—more than cooking pots and storage jars—has a long association with sociability and ritual, status and display. In many cultures, the possession of a particularly fine cup or bowl or dinner service enhances one's standing in the community; often, only the wealthiest members of a society could afford the best dishes.

It is likely that the first pots made only for serving were drinking vessels—tall, straight-sided beakers or shorter, rounded cups. The earliest decorations were fire clouds, black spots from the bonfire that were at first accidental but were later cultivated. These were followed by (and often combined with) simple incised lines: wonderful geometric patterns consisting of crosshatches, horizontal bands, vertical lines, and Xs etched into the damp clay before it dried. These lines made even coarsely formed vessels pleasing to the eye.

Water jar with lovely fire clouds. Low fired pottery is porous and naturally keeps water cool. Note the lugs for carrying ropes.

Fire clouds, as mentioned, were not always considered attractive. Some Neolithic potters worked hard to prevent them. A kiln (and even a bonfire) can have either an oxidizing atmosphere, a reducing atmosphere, or a neutral atmosphere—each of which profoundly affects the look of the pots being fired. In an oxidizing atmosphere there is plenty of oxygen, which, in the heat of the kiln, combines with the metal oxides in the clay body—particularly iron oxide—affecting the color of the finished pot. Red earthenware clay emerges from an oxidizing fire a clear red. In a reducing atmosphere, the supply of oxygen is cut off and the atmosphere is forced to rob oxygen from the iron oxides in the clay being fired. The same red earthenware clay would emerge a black or gray color. In a neutral atmosphere, the red earthenware would remain red, though not as bright as in an oxidizing atmosphere. The oxidizing and reducing atmosphere can take place either as the fire is burning or during the cooling cycle.

A fire cloud is formed when a portion of the pot is reduced. This can happen deliberately, because of where or how the pot was placed in the fire or kiln. Or it could be an accident. Predynastic Egyptians buried the tops of their tall storage jars in a bed of ash inside the fire, thus subjecting only this portion of their wares to reduction. In this way, they created lovely black-rimmed jars with red bodies. In fact, throughout the world many potters discovered independently that they could produce wonderful monochromatic jet black ware by smothering the bonfire in damp straw or dung to cut off the air supply—in other words, by firing in a reducing atmosphere. The Lungshan potters of ancient China made their thinly thrown goblets midnight black by firing them in reduction. The famous black terracotta sculptures of West Bengal are fired in reduction. And the black-on-black pots of Pueblo artists Maria and Julian Martinez of San Ildefonso were fired in reduction.

Another way to impact the appearance of pottery is to use "slips."

Neolithic potters painted designs and pictures on their wares using different colored clays. A potter might gather red clay from a nearby riverbank, yellow clay from a bed that lay a distance away, and gray clay from yet another source. After digging the clay, the potter would carefully pick out the roots, stones, and debris (such as rotting leaves) and then let the clay dry. Once it was thoroughly dry, it could be pulverized and further cleaned before mixing it with water and stirring it to a smooth consistency. This mix of several different colored clays is called a slip. Ancient potters painted robust polychrome designs with these slips. Examples in museums include a seven-thousand-year-old anthropomorphic vase from Turkey, Mesopotamian bowls, Inca stirrup jars, and Minoan jars covered with the swirling tentacles of octopi.

The Greeks brought slip-decorated pottery to its highest level by controlling the atmosphere of their kilns with great precision and by using ultrafine slips made from the same clay as the pots themselves, plus sometimes a whitish slip. The slips were deflocculated—a process for separating the finest particles for use—by mixing the clay with a bit of ash, which caused the heavier particles to sink while the finer particles of clay floated on top. They skimmed the resulting creamy slip from the surface, and mixed it with urine or wine for use. In the earliest days, they painted intricate geometric designs. Later they wrapped their pots in scenes from daily life, including sexual escapades, exercise workouts, feasts, celebrations, musical performances, myths, and even pot making.

The Greeks developed an amazing firing schedule and demonstrated a deep understanding of reduction and oxidation. By controlling the atmosphere of the kiln, they were able to produce beautiful, two-colored pots.

The early hours of the fire were in oxidation, up to 1,650°F (900°C). If pulled from the kiln at this point an entire pot, including the design, would be red. However, the potters then threw damp wood or sawdust, perhaps even a bit of water, into the kiln. Using a long tool, they reached up to the top of the hot kiln and closed down the damper, at least partially. The kiln was now in reduction. If pulled from the fire at this point, pots would emerge black on black. Instead—and here is the genius—the potters would briefly oxidize by introducing tinder-dry fuel and opening the damper. This created enough of an oxidation atmosphere to turn the pot back to red, but not enough to oxidize the slip, which remained black. This could also be done in reverse, with the background done with the slip.

The slip was not a glaze as we define a true glaze today; it did not form a glassy coating high in silica. It was, however, glossy and vitreous and, in practice, functioned as a glaze. Using this ingenious firing method the Greeks produced their spectacular wine cups, vases, serving pieces, and other tableware—plus pieces for show. Ownership of such fine wares lent prestige to a household. It was no longer enough that a drinking vessel be serviceable; now, if you were anyone, the dishes you used, at least for entertaining, must also be beautiful.

The Etruscans (and then the Romans) were content with glossy black tableware. They were not interested in covering their pots with illustrations. They likely learned how to make deflocculated vitreous slips and the art of reduction firing from the Greeks who had inhabited parts of the Italian peninsula. The Romans may even have hired Greek potters to work in their potteries. And then, for some reason—to save fuel, to meet the demands of fashion, for ease of firing—Roman potters ceased reduction firing and fired in oxidation. The ware, covered with the same fine slip, now came from the kiln a bright, glossy red rather than black. Roman pot shops made bowls, one- and two-handled cups, beakers, ewers, stemmed cups, dishes or platters, bottles, and of course amphorae.

Roman tableware was thrown, molded, sprigged, and stamped. The molds, made of fired clay, were centered on the wheel. The potter then pressed the clay into the mold, against the walls. The resulting beaker or bottle could be quickly made in this way with an elaborate raised design. A Roman monochrome red cup with highly embossed images had a very different look from hand painted Greekware. They could be mass-produced and were easier to fire, yet they exhibited an unassuming and honest beauty.

The finest pots were made in Arretium (modern Arezzo) from about 30 B.C.E. to 30 C.E., and for that reason these pots are sometimes called Arretine ware. They are also known as *terra sigillata*, or "earthenware decorated with small figures."

Romans also produced simple coarse wares for cooking and for the ordinary household with meager financial means. These pots were left plain, without a coating of slip, and without relief designs. They were quickly thrown on the wheel and often had strong shapes, but they were not valued as highly as terra sigillata. The Romans set up pottery workshops throughout their burgeoning empire and exported their methods of working. Usually, Roman-style pots replaced local styles and methods.

It was glazed ware, however, that most enhanced the dining experience.

A glaze is a layer of glass that coats the pot, making it waterproof, stronger, and often very beautiful. Today, the formulation of glazes is a science utilizing chemical formulas and calculations, though test fires are still necessary. The kiln fire brings mystery and surprise— a glaze might have a perfect formula yet look boring or ugly—how it will appear can only be discovered by doing a test-firing.

We know the chemical composition of the various components, and in what percentage they appear in the glaze. At its most basic level, this is expressed as $RO\ R_2O_3\ RO_2$. A glaze needs a flux to cause the silica to melt (expressed as RO) a refractory element such as alumina to give the glaze durability (expressed as R_2O_3), and a glass former, silica (expressed as RO_2). In reality, however, a glaze formula (or recipe) is never quite so pure and simple, because the various components are derived from a combination of ingredients.

For instance, the silica may come from both sand (quartz) and from clay, which contains silica.

At peak temperatures, glazes are molten in the kiln. If a glaze is fired to too high a temperature, it will drip, run off the pot, land on other pots, and cause a kiln disaster. The bottoms of pots must be cleaned of all glaze before being placed in the fire, or they will stick to whatever they touch—another pot, the kiln floor, kiln furniture such as a saggar (refractory box used to keep the ware from contact with the flames), or a shelf. Potters must also be careful to keep pots from touching each other lest they stick.

Glazes are applied to pots by dipping, pouring, spraying, dry dusting, and painting. They can be applied over one another, scratched through to the clay to reveal a design, or decorated with oxides. The possibilities are endless.

The Egyptians, and perhaps Mesopotamians before them, developed "Egyptian paste," a self-glazing body, by 4000 B.C.E. The salts in the body migrate to the surface during the fire and turn a brilliant turquoise blue if copper is present, purple if manganese is present. Egyptian paste, sometimes incorrectly called Egyptian faience, was shaped in molds and by hand and is used by artists to make beads and pendants today. It was not plastic enough to be thrown or coiled.[1]

Low-fire glazes are divided into two types, categorized by the fluxing agent: lead and alkaline. Both types have been known since at least 3000 B.C.E. The Egyptians discovered glass making and by 3000 B.C.E. they knew how to make a true glaze, but generally limited its use to bricks, tiles, and wall plaques.

Lead is a powerful flux, melting at 950° F (510° C). It rarely occurs as a pure ore, but is found throughout the world as galena, red lead, white lead, lead carbonate, and other compounds. Lead glazes have a wide firing range (which is very helpful to potters relying on uneven updraft kilns), so overfiring and underfiring does not lead to the daunting problems presented by other types of glazes. Lead glazes also take color well, can be transparent or opaque (depending upon the additives), flow evenly rather than run, and resist crazing (crackling). Lead glazes have been popular for thousands of years. The warm honey color of English country pottery, the soft orange-red of rustic early American tableware, the golden brown medieval jugs and bowls that appear in paintings, unsurpassed in their forthright, honest beauty, obtain their appealing clarity of color from the lead glaze that adorns them.

Unfortunately, lead is highly toxic. A potter mixing a lead glaze could easily inhale a cloud of lead dust and endanger himself, or inadvertently ingest some glaze caked on his fingers during a lunch break. And even after firing, when a lead-glazed piece of pottery is used to serve food, danger still remains: acidic foods and beverages leach the toxic substance from the glaze, exposing the diner to lead poisoning. Symptoms of lead poisoning vary from headaches and nausea to brain, kidney, or nervous system damage or failure, and sometimes death. The ancient Greeks suspected the toxic nature of lead, but it was centuries before anyone knew for certain that lead was a poison.

We do not know when lead glazes were first made, but archaeologists have recovered lead-glazed pottery from Asia Minor, Mesopotamia, and Egypt, where it was used as early as 100 B.C.E.[2] It was added to alkaline glazes as early as 1000 B.C.E. but only in minute quantities to alter the color.[3]

Lead glazes were used in Greece and, later, Rome. Romans produced pretty green and yellow lead-glazed ware for centuries, but they preferred their red slipped tableware until the reign of Constantine, in the first decades of the fourth century C.E., when lead-glazed wares began to dominate. With the collapse of the Roman empire, many skills and technologies were lost, but potters continued to make humble, lead-glazed dishes, platters, cups, jugs, and flasks to meet the needs of their communities. When invading marauders from central Asia invaded China, plunging the continent into the dark period between the Han and Tang dynasties, the art of lead glazing was lost. It was not rediscovered until trade resumed in the Tang period (618–907 C.E.) and, as in the past, it was used more for figurines and pots for the grave than for tableware.

Alkaline glazes, though safer and more ancient, were more problematic and unstable to use. They could produce wonderful blues and greens, but reacted harshly with the raw clay of the pots. It would be centuries before suitable techniques and formulas were developed for them to be in common use.

The great innovations in glazes—innovations that changed the way potters in distant Europe worked and in the tableware that kings and peasants dined from, and in the way we set the table today—came from China, where the roaring infernos of climbing kilns spewed wood ashes on the ware stacked inside. At high enough temperatures, wood ash melts into a natural glaze. Shiny, mottled green and amber-colored natural ash glazes, known as "kiln glost," formed on the shoulders and rims of ancient Chinese pots.

But ash deposits are unpredictable and difficult to control. It would be natural for potters to attempt to find a way to reliably reproduce their best results. The easiest course would be to sprinkle ashes onto the unfired pot before it was placed in the kiln, thereby assuring that it would be glazed. If one wanted the interior to be glazed, why not mix the ashes with a clay slip so that the ashes would adhere, and slosh the mixture around inside the pot until it was coated? And if you are going to mix clay and ashes, why not try some ground-up stone, such as feldspar?

Chinese potters were firing at stoneware temperatures and producing watertight vessels before they made glazes. Unlike earthenware, stoneware does not need a glaze to be watertight. It was the attractive appearance of glazed ware, and perhaps the smooth feel of a glazed pot when it's held in the hands or touches the lips, that the Chinese liked. Glazes add to the aesthetics of tableware.

Through the centuries, the vast lands of China were disunited, united, conquered, invaded, disunited, and united again under various rulers. Pottery traditions and skills sometimes developed independently in the north and the south and varied from one province to the other, but more often, word of new methods traveled and fashions spread. Even in the most turbulent times, Chinese peoples, and their conquerors, appreciated and used the bowls, cups, ewers, plates, and dishes that came from the many kilns that dotted the land. The work of the best kilns was frequently reserved for the rulers, the kilns and workshops sometimes owned by the rulers, but common people also had access to and appreciated fine dinnerware.

During the reign of the Shang clan (1751–1111 B.C.E.) on the northern plains of China, arts and crafts flourished. Li Zhiyan and Cheng Wen note that there were "six handicraft trades: potters, smiths, stonemasons, carpenters, animal tenders, and straw weavers. Potters, who made tiles as well as vessels, ranked first."[4] Potters produced thinly glazed pots fired to 1,980 °F (1,080 °C) or higher. The glazes appear to be made of ash mixed with lime or earth and were greenish yellow, with spots of deeper green.

The Shang also made ceremonial wine cups, urns, and ewers from a white clay that was primarily kaolin. It was stiff and not easily worked, and was fired to about 1,832 °F (1,000 °C), which left it somewhat fragile. This was not the first appearance of white ware in China—the Dawenkou culture of 3000 B.C.E. had produced white pottery—but the Shang pieces were well made and highly valued.[5]

The Shang tribe, according to legend (though the famous historian Sima Qian [145–86 B.C.E.] claimed this story was the absolute truth), sprang into being when the Emperor's wife gave birth to a remarkable baby after she accidentally swallowed a bird's egg while taking her bath. One wonders how this was possible. Was she bathing outdoors, stretched out in a tub with her mouth open, when an egg fell from a nest in a nearby tree? Or did a swallow swoop down and drop an egg into her mouth, as one poet claimed? Who knows, but the improbable boy supposedly resulting from this odd occurrence grew up to become the mighty ruler known as Lord of the Shang. He and his descendants remained in power for six hundred years until they lost their power to the invading Chou, who, beginning with the Emperor Wu, reigned from 1155 to 255 B.C.E. Wu was famous for his appreciation of a good pot.[6]

Specialist Margaret Medley believes that "the evolution of glazed pottery was not an uninterrupted development, and the intermittent occurrence, without apparent connection in various parts of China over the succeeding seven or eight centuries, makes the plotting of any steady growth in glazing skill exceptionally difficult."[7]

Nevertheless, at least some Chou dynasty pots seem to be directly derived from the Shang. Chou potters were able to produce true stoneware glazes that they fired to at least 2,192 °F (1,200 °C). These glazes were thicker than the Shang glazes and more even.

Both the Shang and the Chou had specialized pottery workshops, some producing only one type of vessel in vast quantities for market. A skilled potter was held in high esteem. *Zuo Qiuming's Chronicles* tells us "Yu Efu was a potter in the early Zhou Dynasty. King Wu relied on him to make good pottery vessels. After Yu became a wondrous craftsman, the king took him as son-in-law and invested him with the State of Chen."[8] A province of your own is quite a reward for making good dishes. Hopefully, Wu's daughter appreciated the potter and the pots. The Shang and Chou produced wine goblets, urns, ewers, bowls of various shapes and sizes, jars, steamers, and a multitude of cooking and storage vessels.

During the Chou dynasties, both Confucianism and Taoism were born. Confucius (551–479 B.C.E.) traveled widely, teaching harmony in thought and deed and reverence for parents and ancestors. Lao-tzu, about whom we know very little, is believed to have founded Taoism, which embraces an acceptance of what is and simplicity in one's life. Both philosophies had a deep influence on the people and culture of the Chou era. Perhaps the most direct influence on clay workers was Confucius' repudiation of the then accepted custom of interring a highborn man's wife and slaves with him at his death; his followers used clay—and on occasion wood—substitutes instead.

Han dynasty (206 B.C.E.–220 C.E.) potters made quantities of high-quality green or greenish yellow glazed stoneware. They had efficient wheels and improved kilns. During this period, both Confucianism and Taoism took even deeper root and, in the first century C.E., Buddhist missionaries from India introduced their religion to China. The Han expanded their territory into Korea, built roads, fended off invading "barbarians," and widely extended their trade, especially with the Near East, Persia, India, central Asia, and the eastern Roman empire. It was perhaps through this trade, when they were exposed to information and ideas from great distances, that they learned the art of lead glazing earthenware from Egyptian or Persian potters. We know that they imported a lead "frit" (partially fired lead, which was easier to transport and use) from the Near East. There is no way to know for certain, however, whether the Han developed the process independently or imported it, but it is reasonable to guess that they might have learned it from contact with people who had been utilizing lead glazes for a thousand years and who had more lead at their disposal. The Han, however, restricted the use of lead glaze to funerary ware and, fortuitously, did not use it on vessels for eating.

The Han dynasty collapsed with the invasion of central Asian warriors. China plunged into its "dark ages," called the "Six Dynasties" or "Dynasties of the North and South." Years of disunity, anarchy, and fractious wars ensued. China was split apart and there were repeated but short-lived and bloody attempts to reunite it. Nomadic invaders crossed the old borders and waves of "barbarians" entered from the north. Trade was hampered though it did not cease. Knowledge of lead glazing was lost and did not reappear for centuries, but throughout these troubled centuries, potters continued to produce glazed stoneware.

In 581 C.E. the Sui dynasty, founded by the aristocratic Yang family who had taken control of northern China, installed Wendi as emperor and after eight years, they had reunited with the south. The dynasty was short-lived, ending with the death of Wendi's son in 617 C.E., but during its brief tenure there was an influx of foreign goods and stylistic influence. The Yang family was particularly fond of items from Persia.

Potters of the Sui dynasty, influenced by the new fashion for Persian imports, rediscovered the lost art of lead-glazed pottery. They used multiple glazes on a single pot, coloring one glaze with copper for green, another with iron oxide for honey brown. They continued to make high-fired stoneware. Some of their clay bodies were highly porcelaneous and were covered with a smooth glossy white or light gray glaze. Although it is impossible to know when the first porcelain was made, many experts consider this light-colored Sui pottery the first example.

Chinese potters had been interested in white wares for centuries. They had also learned to mix different clays from distant mines to make a body that could be easily thrown on the wheel. Clays that are hand or coil built can contain some impurities, but for throwing, clay must be smooth and plastic. A pebble in the wall of a spinning pot will throw the pot off center or tear the wall as the clay revolves in the potter's hands. Sandy clay is abrasive and unpleasant, even painful, to the thrower's hands.

The Chinese call any high-fired clay body that makes a clear ringing sound when it is struck *ci* (*tz'u*) or porcelain. In the West, a body is considered true porcelain if, in addition to these criteria, it is white and translucent. Porcelain is made of kaolin (white china clay), feldspar, and silica. Kaolin is named for the hill of Gaoling (Kao-ling), which lies to the north of Jingdezhen (Ching-tê-Chên). Kaolin is 40 percent alumina, 46 percent silica, and 14 percent water. It is highly refractory and pure white. Petuntse (china stone or literally, "small white rocks") is a naturally occurring white, feldspathic powdery rock that occurs in China. It was added to kaolin to make porcelain. In the fires of a very hot kiln, petuntse melts and surrounds the refractory particles of kaolin, giving the body strength and making it smooth, almost glassy. Porcelain is fired to 2,280–2,370°F (1,250–1,300°C).

It was during the Tang dynasty 618–907 C.E. that the art of porcelain was perfected and in the following dynasties it came to dominate. The three centuries of the Tang dynasty were a time of creativity and prosperity. The borders were extended to include Korea, Tibet, Annam (Vietnam), and eastern Turkestan and were relatively secure. Extensive international trade was resumed and an influx of immigrants, including "Persians, Turks, Uighurs, Tibetans, Sogdians, Indians, Greeks, Koreans, Japanese, and people from Khotan and Jucha" brought goods and ideas and commerce.[9] Buddhism was now widely practiced, with the heavily populated city of Chang'an (Ch'ang-an) becoming "a great center of Buddhist culture."[10]

In the past, craftsmen were required to produce for the royal court, but now they were monetarily taxed instead, and a new class of independent hand workers arose. Crafts and the decorative arts burgeoned. Poetry thrived and tea drinking became an art form. Tea was cultivated in the southwestern parts of China by the third century of the Common Era, but during the Tang dynasty the cultivation, preparation, and consumption of tea along with a connoisseurship of the fine cups in which it was served defined the culture.

The Tang dynasty is known for four types of pottery: so-called three-color ware (low-fired, lead glazed); luscious celadons; white on white; and, to a lesser extent, black, or tenmoku. The ceramic technology that the Tang workshops developed to create these wares was the most advanced in the world at the time, and was unknown outside Asia until centuries later.

A multitude of goods and ideas flowed into China, including well-made lead frits from mines in the Near East. Tang potters added powdered iron, copper, antimony, and manganese to their lead glazes to make their famous three-color ware. In the clean, oxidizing fires of their kilns these metallic oxides turned the pots brilliant browns, greens, yellows, and purples. Cobalt, which was not discovered in China until hundreds of years later, was imported in small amounts from the Islamic world, where it was used to produce a brilliant blue. It was called Mohammedan blue. Often, the earthenware vases and bowls were dipped into a light-colored clay slip before glazing, which made the colors even brighter. Tang three-color ware in reality was often more than three colors, the various metallic oxides melting, mingling along the edges to make new colors, splashed or applied as dots, the bright greens and deep blues and clear yellows like nothing that had been seen before.

Along with this bold low-fired three-color ware Tang workshops produced exquisite monochrome high-temperature work in green, black, and white. At high temperatures, in a reducing (smoky) atmosphere, iron produces a green glaze known as celadon, rather than the red or brown of an oxidizing fire. The Chinese had been making green or greenish-colored ware for centuries. Now they used a light-colored clay body, frequently porcelain, which clarified the color of the glaze. The green was soft, jadelike, the tint of a spring apple, a pool of mossy water, on occasion even misty bluish. The contemporaneous poet Xu Yin attempted to describe a teacup in this magical hue in his poem "Secret Colour Porcelain Teacup, a Leftover Tribute Piece":

> Blend green and blue into a refreshing hue,
> We offer new porcelain as tribute to the throne:

> The cups intriguingly fashioned like the full moon softened by
>   spring water,
> Light as thin ice, setting off green tea,
> Like the ancient mirror and dappled moss placed on a table,
> Like dewy budding lotus blooms bidding the lake farewell,
> Green as the fresh brew of Zhongshan's bamboo leaves,
> How can I drink my fill, ill and frail as I am?[11]

According to Lu Yu, the eighth-century tea aficionado, who wrote the *Canon of Tea* in which he rated the many kilns, the celadon tea bowls of Yuezhou were superior to all others. Indeed the best of the Yue celadon was reserved exclusively for the rulers. Again, here is a contemporaneous poet, this time the ninth-century, Lu Guimeng, describing "Mi Se," or secret color:

> In full autumn wind and dew the kilns of Yue open:
> A thousand peaks are despoiled of their halcyon green.
> Let us fill Yue cups at midnight with wine like nectar
>   from the sky;
> Let us drink with the poet Ji Kang whose capacity vies
>   with his verse.[12]

The Tang, especially the ruling class, appreciated beautiful dishes for their meals. The most perfect tea bowls and wine cups from the best kilns were given as tribute. Production of white glazed porcelain, sometimes more vanilla colored, or with a grayish tint, also came into prominence during the Tang dynasty. It was popular with both common people and the upper classes. From the poet Bai Juyji:

> We eat rice and celery,
> Using white porcelain bowls and green bamboo chopsticks.[13]

One almost expects to see a glossy magazine photo or still life in oils of this place setting: the white of the rice contrasted with the light green of the celery, the white celery hearts, the satiny white bowl, the green chopsticks for eating, a harmonious balance of color and texture. The Tang epicures knew that a white dish never visually competes with the food. Instead it acts much as a frame does for a painting, showing off the meal, enhancing it.

■

THE CHAOTIC AND tumultuous Five Dynasties and Ten Kingdoms (907–979 C.E.) followed the Tang dynasty. The Song (Sung) dynasty (960–1279 C.E) took control of the north in 960 C.E. and by 979 C.E, had reunited the north and the south of China; however, China split apart again when fierce invaders from the windswept steppes of Manchuria conquered the north, leaving the Songs only the south.

The three centuries of Song rule fostered advances in science and technology, education, manufacturing, and the first use of paper money in the world. Frequently cut off from trade, and constantly threatened by encroaching armies, the Song still produced the best ceramics the world has known. Workshops turned out an abundance of true porcelain, as much as ten thousand teacups in a kiln load! The smooth, fat celadons of the Song surpassed those of the Tang dynasty. In the north, even after the Song no longer ruled there, the remaining potters kept the Song kilns burning and made quietly beautiful dishes, cups, and bowls.

During the Song dynasty, workshops specialized in making a few items in large quantities. Highly skilled artists carved intricate flowers, birds, grasses, fish, ducks, and sometimes dragons into walls of the masterfully thrown leather hard pots. Potters began the practice of twice firing some of their work, first at a lower, or "bisque," temperature, followed by a second, high-temperature glaze firing (a practice followed by most studio potters today). This was done because it was easier to glaze a fired pot than a fragile, unfired pot, and because some glazes only worked on a fired pot. Demand was high and eventually workshops blossomed into factories. New methods of making pots were invented. Carving required a very talented and skilled hand and was time consuming. Potters continued to throw on the wheel, but now they also used molds. With molds, elaborate designs could be duplicated in quantity with less dependence on skill.

Thick rounded or humped molds were thrown on the wheel. An artist then incised the outlines of peonies, chrysanthemums, cherubs, terraced gardens, fish, birds, or vines with sharp, deep cuts in the convex side of the mold, which, when dry, was fired almost to stoneware temperatures. To use a mold, a disk of clay was placed over it and beaten with a paddle, a task that was easily mastered. In this way, multiples could be efficiently made.

Soon an even more efficient method was devised; the mold was fastened to the wheel and covered with a thick disk of clay. As the wheel was spun, the potter held a template against the clay. With the template in position, the clay was simultaneously pressed into the carved mold to form the interior of the pot, while the exterior was scraped into shape by the template. In modern factories today similar devices ("jiggers" and "jolleys") are widely used to mass-produce dinnerware.

In some instances, molds were used to alter the shape of a thrown pot. "Sprigs"–small molded clay decorations, like leaves–were also added to Song pottery. Today, sprigs are most commonly used on Italian flowerpots.

Bowls, plates, teacups, and dishes, thrown or molded, were also trimmed to make the walls thinner than possible during throwing and to cut a foot ring. This meant returning them to the wheel and, after setting the wheel in motion, using a sharp tool to shave the walls and carve a foot ring in the base. In this way the Song made exquisitely translucent thin-walled porcelain vessels.

Song workshops excelled in firing techniques, methods of forming, carving, design, and glaze making. Glazes were quiet and subdued in appearance, yet, if you held a teacup in your hand or gave a bowl more than a glance, you would notice subtle variations. The longer you looked, the more you saw. The "tear stain" where the glaze ran down toward the foot, leaving a thicker coating of glaze, the side of the pot licked by the flames, the shoulder where the glaze thinned, showing the color of the clay body beneath, all added to the visual interest of the vessel. There were shimmering celadons, silky whites, iron-rich dark tenmokus (sometimes with the shadow of a single, saturated golden leaf fired into the interior of a bowl), and the legendary opalescent chuns (jun) with tints of blue and purple and red. In chuns, bubbles are suspended inside the fired glaze, thus bending and refracting the light, making the eye see blue, though there are indeed no blue colorants in the glaze. Reds, and purple flashes, came from the copper.

There is an old Chinese legend that has been told and retold by potters far beyond the Chinese borders as well as by generations of Chinese. Of course, with so much retelling, there are variations, but the basic story remains the same:

One day, a potter unstacked his kiln and found deep inside, amid all the other pots, one perfect red pot, like nothing he had ever seen before. It was a beautiful deep shade of crimson from the neck to the gently curved belly to the delicate foot, with just a bit of white at the rim. Knowing how much the emperor enjoyed a fine pot, he sent it to the palace. Indeed, the emperor was overjoyed and immediately placed an order for a thousand red pots. He must be the first to possess such amazing objects of breathtaking beauty.

The potter despaired. He did not know why the pot was red. But the emperor must not be disappointed, and so the potter set about the task of making more red pots. He mixed up his finest porcelain clay and prepared it well and then threw pot after pot and glazed them with his best glazes. He split the driest wood that he could find, and carefully stacked his kiln and fired his pots, hoping that he could reproduce the miracle of the red pot. But a few days later, when he opened his kiln, he could not find a single red pot. He worked day and night to make and fire more pots but each time he opened his kiln, he grew more depressed. The emperor grew impatient and sent word that he wanted his order delivered or else! One more time the potter made enough pots to fill his great kiln. But why would this fire be different from the others? He paced back and forth past the mouth of the roaring kiln and wrung his hands with worry. As the kiln temperatures climbed he grew more fretful, and finally, fearing he would fail once again and incur the emperor's wrath, he jumped into the kiln himself despite frantic pleas from his helpers. He disappeared in the flames.

When the kiln was cool enough to open, his helpers unbricked it and were astonished to see thousands of perfect red pots!

This legend explains the origins of copper red glaze, which the Song perfected.

In some versions, a pig or mouse gets into the first kiln load with the red pot, unbeknownst to the potter. Though just a story, there is some science behind it: happily, red glazes do not require self-immolation, just a reducing fire. In the same way that iron turns green in reduction, copper (which is green in oxidation) is red in reduction. These types of glazes are a bit more troublesome than other glazes and reduction must be at just the right moment and to the right extent. This difficulty probably gave rise to the story, which neatly conveys the Chinese potter's willingness to experiment to reproduce results, and the deeply ingrained appreciation of pottery in the culture.

Tea drinking became even more important during the Song era than in the preceding eras.

Plants were now carefully tended, the leaves were picked at precisely the right moment, and the tea was pressed into cakes, which were shaved as needed. Contests were held at court and participants were judged on their ability to discern differences between teas. The Song paid particular attention to the aesthetics of tea drinking, and liked their teacups to enhance the tea's rich, rosy color. Tea drinking became entwined with Buddhism, both of which spread to Japan, where the tea ceremony evolved.

The Song dynasty eventually collapsed, even though it paid tributes of tens of thousands of pots to the threatening hordes beyond their borders. As part of the Mongol invasion that ultimately overran the Song, Kublai Khan established the Yuan dynasty in 1260 C.E., which ruled a united but conquered China for 108 years.

■

UNDER MONGOLIAN RULE, China's international commerce expanded. Impressed with the popular blue and white earthenware that merchants brought from the Middle and Near East, potters under the Yuan dynasty began to make their own version of blue and white ceramics. Using imported cobalt, powdered and mixed with water, they brushed designs onto their porcelain vases and dishes before glazing. Once an impure form of cobalt was discovered in China, they mixed that with the more expensive imported oxide.

It was during this time that Marco Polo traveled with his father and uncle to visit Kublai Khan. He wrote, "Of this place there is nothing further to be observed, than that cups or bowls and dishes of porcelainware are there manufactured. The process was explained as follows. They collect a certain kind of earth, as it were, from a mine, and laying it in a great heap, suffer it to be exposed to the wind, the rain, and the sun, for thirty or forty years, during which time it is never disturbed. By this it becomes refined and fit for being wrought into the vessels above mentioned. Such colours as may be thought proper are then laid on, and the ware is afterwards baked in ovens or furnaces. Those persons, therefore, who cause the earth to be dug, collect it for their children and grandchildren. Great quantities of the manufacture are sold in the city, and for a Venetian groat, you may purchase eight porcelain cups."[14] Curiously, in Italian, *porcellane* means "little pig" or "cowrie shell." Marco Polo may have been the first European to bring porcelain to that continent.

In the ensuing years many craft workers, including potters, were reduced to serfdom. There were sporadic uprisings and revolts. The plague arrived in 1330 followed by a string of floods and famines and more unrest. Finally, in 1368, the Chinese managed to throw off the yoke of the Mongols, and established the Ming dynasty (1368–1644).

For the first time in four centuries, Chinese ruled China in its entirety.

The city of Jingdezhen grew into a bustling center of pottery making. The Ming workshops continued the traditions of stoneware in addition to porcelain, and carried on production of celadon, chun, tenmoku, and pure white porcelain as well as special pieces in copper red. They also produced a type of "secret" white porcelain, with birds and blossoms painted in white slip under a clear glaze or lightly carved through the glaze to the body so that the design could only be seen when the pot was held up to the light.

Though they appreciated glaze subtleties, the Ming were not as devoted to monochromatic ware as the Song. They loved color; color in their wardrobes, their homes, their gardens, and in their ceramics. In answer to this, potters developed a technology for making high-fired stoneware and porcelain with decorations in colors as brilliant as those of Tang low-fired earthenware. To achieve this, they glazed and fired the ware to high temperatures in the traditional solid shades. They then painted elaborate designs, plum blossoms, gardens, dragons, pavilions, lotus blossoms, phoenixes, scrolls, and such with oxides and lead glazes onto the fired glaze and refired the pots at lower temperatures in special muffle kilns. These new kilns were built so that the ware was stacked in a large inner chamber beyond the reach of the flames. This new process was a dramatic technological breakthrough. To collectors today, these bold, multicolored pots with elaborate allover designs are spectacularly gaudy or richly gorgeous, depending upon one's sensibilities. They were wildly popular at the time and admired wherever merchants transported them.

Early in the sixteenth century, the Ming developed a special high-fire, glossy red clay that did not require glazing and used it to produce teapots. Some of these pots had the startling quality of changing color when hot tea was brewed in them. R. L. Hobson explains, "Yi-hsing Hsien (Ihing) is situated on the west of the Great Lake in Kiangsu. The materials miraculously discovered in the neighboring hills include a variety of clays which, after firing, produced wares of cinnabar red, dark brown, 'pear-skin,' green and light red colours, while other shades could be obtained by discreet blending. A priest of the neighbouring Chin-sha temple was the first person to put these clays to their proper use, viz, to make 'choice utensils for tea-drinking purposes,' and his secret was surreptitiously acquired by one Kung Ch'un, whose subsequent fame far surpassed that of the true originator of the ware. Kung Ch'un lived in the Chêng Tê period (1506–21), and his teapots are described as 'hand made with thumb-marks faintly visible on most of them.' They were chestnut coloured with the 'subdued luster of oxidized gold,' and their simplicity and accuracy of shape were inimitable, worthy, in fact, to be ascribed to divine revelations. Hsiang's Album purports to illustrate two teapots by Kung Ch'un which were bought for five hundred taels. One is hexigonel and of grayish brown colour like felt, and the other ewer-shaped and vermilion red, and both had the miraculous quality of turning to a jade green colour when the tea was made in them. Hsiang adds that if he had not witnessed this phenomenon with his own eyes he would not have believed it."[15]

We can guess that the clay bodies contained titanium dioxide and rutile either naturally or as additives. Titanium dioxide is a fairly common mineral, accounting for 1 percent of the earth's crust. It often occurs in sand. Rutile is an iron-rich version of titanium dioxide. Titanium dioxide used in pottery has two strange qualities: phototrophy and thermotrophy. Phototrophy is its ability to bend light in a glaze. It is this quality that gave the chun glazes their quality of looking blue without any blue colorants. More startling is thermotrophy. Certain combinations of titanium dioxide with small amounts of rutile, in a glaze or clay body, look white at room temperature but appear yellow at temperatures over 302 °F (150 °C). This is probably what made the red teapot turn green once the hot water was poured in.

The scholarly and literary classes during the Ming era particularly favored Yixing (Yi-hsing) teapots, which came in several shapes. They eschewed the more ostentatious, brightly colored teapots.

Later, when the tea craze swept through Europe, red Yixing teapots were particularly in demand, especially by the wealthier classes. By the waning years of the seventeenth century, China was exporting large quantities of these pots, though they were no longer noted for changing hue.

But it is the wonderful calligraphic blue and white porcelain dishes for which the Ming dynasty is best known and which were most popular in China and, as they were exported, throughout the Western world. Skilled artists dipped their brushes in cobalt and painted blossoms, birds, acrobats, magnolia trees, lovers, flowering branches, pavilions, and swirling vines and scrolls on porcelain dishes, teapots, bowls, teacups, vases, jars, ewers, and plates. There is something mysterious about the deep rich blue of cobalt together with the arctic white of porcelain that elicits a passionately positive response in all but a very few who behold the combination. Blue and white evoke a sense of serenity, purity, and, perhaps royalty in the user or viewer. If you go to a craft show today and talk with exhibitors you will soon hear a potter mention (or complain) that all customers love blue. Similarly, wherever Ming blue and white porcelain was exported, it became highly favored. In Sumatra, Egypt, Persia, and the Near and Middle East, Ming blue and white porcelain was highly sought after. By the time of the Qing dynasty (1644–1912) Chinese potters were producing and exporting enormous quantities of porcelain, predominantly blue and white, often tailoring it to their client's needs. It is this huge porcelain export business that gave dishes or tableware the name of "china." We serve our guests on our best china. We keep our dishes in a china cabinet. We worry about a bull in a china shop. During this time, matched dinner services came into favor. By the middle of the sixteenth century they were de rigueur. One official royal household inventory list from 1544 accounts for 1,340 sets that "comprised twenty-seven pieces, five fruit dishes, five food dishes, five bowls, five vegetable dishes, three tea cups, one wine cup, one wine saucer, one slop bowl and one vinegar cruet."[16]

Of course, local potters in the Middle East, central Asia, and Europe were not happy with the competition from the Ming potters, and they attempted to produce similar ware themselves. But they had neither the proper clay body made of kaolin and petuntse nor kilns that could reach high enough temperatures.

What they did have, however, was motivation and tin oxide.

During the ninth century Tang dynasty potters were so besotted with the beautiful blue glazes that the Islamic potters of the Baghdad area were able to produce they paid handsome sums to import cobalt oxide to use themselves. At the same time, the Islamic potters were very taken with the milky white porcelains that came from some of the best Tang kilns, but they were unable to replicate it.

However, in their efforts to imitate porcelain, they discovered that the addition of tin oxide in amounts of up to 15 percent, makes a glaze opaque and produces a smooth, creamy white skin. It is a perfect background for decoration.

For centuries, Chinese potters had kept the secrets of porcelain. The art of tin glazing, however, spread from the Islamic world to Moorish Spain, Portugal, France, and Italy and on to the Netherlands. It developed into the rich traditions of lusterware, multicolored maiolica, and faience, and as miles and years added distance, tin glazing departed from the original Chinese influence (an attempt to imitate porcelain), and became an oeuvre in its own right. Then the Dutch went to China and everything changed.

Guido di Savino, a friend of the Italian Cipriano Piccolpasso who described the art of maiolica glazes in *The Three Books of the Potter's Art*, set up an Italian style maiolica workshop in Antwerp around 1512. At first the workshop produced brightly decorated drug jars to hold spices and pharmaceuticals, and tiles, both in blues, yellows, greens, and oranges. But after the Dutch adventurers reached China in 1596, and they beheld the lovely blue and white porcelains of the Ming dynasty, and then six years later the Dutch East India Company began importing these much-coveted dishes into Holland, the workshop turned to producing faux Ming ware. Potters faithfully copied the Chinese motifs at first but soon they inserted symbols from their own culture, mixing the two together in sometimes surprising ways, juxtaposing Buddhist and Christian symbols on the same bowl. Delftware, as it was called, became hugely popular and soon became an export itself.

During the many years that the Chinese were writing poetry about their lovely celadon teacups and dining from delicate porcelain bowls, Europeans were content with their rustic, lead-glazed earthenware in sturdy utilitarian shapes. Often, they drank from wooden tankards or ate their meal off a slab of bread. Pottery bowls and jugs and mugs were thickly thrown and had a simple, honest beauty to them. A diner might have his or her own cup but would share a mess or platter with others at the table.

A glimpse at the inside of the oven, where the ware is fired, showing how the saggars are built up in bungs.

Peasants continued to rely on these simple earthy dishes for centuries. But the aristocrats preferred the new tin-glazed dishes. These were so much easier to care for than tableware made of metal, were pleasant to eat from, and did not tie up the family's wealth. By the beginning of the sixteenth century, ceramic tableware was preferred by everyone. Soon, influenced by Ming imports, aristocrats required matching sets. By the middle of the century, a wealthy household might boast a dinner service consisting of two to four hundred pieces. Dinner sets became the norm. In Italy, expectant mothers were given special sets of dishes specifically designed for feeding a newborn. The arrival of Chinese porcelain in Europe created a frenzy in the marketplace and changed the dinner table.

In the Rhineland area of Germany, potters were fortunate to have beds of plastic gray clay that could withstand high temperatures.

They began using this stoneware clay possibly as early as 1000 C.E. but certainly by the twelfth to fourteenth centuries. The area was thickly wooded and provided an abundance of fuel, making the high temperatures at which stoneware matures achievable. German potters turned out tall beer mugs, wine bottles, and storage jars. With the Rhine River nearby, and an already active river trade, markets for the stoneware pots were soon opened up throughout the country and beyond. And then, sometime between the end of the fourteenth century and the beginning of the fifteenth, something unexpected happened: Germans began throwing salt into their kilns!

There is a story that a potter about this time neglected to fill his woodshed prior to firing his kiln. He thought he had enough fuel for one more kiln load before he had to go out into the forest with his ax to fell more trees. However, as his kiln approached top temperature, he realized that his woodpile was depleted. He was desperate. If he shut down now, the beer mugs that filled his kiln would all have to be fired again, a waste of work and fuel. Frantically, he looked around his workshop and kiln shed and noticed an old sauerkraut barrel. Quickly, he rolled the barrel to the firebox and fed it into the fire, one stave at a time. The next evening, when he opened the kiln, he saw that each of the pots glistened as if magically coated in a crust of ice. Thus salt glazing was invented.

Of course, this may be just a myth, but German potters *did* discover that salt makes a wonderful glaze. At temperatures of 2,192 °F (1,200 °C), the sodium in salt chemically combines with the surface clay of the pot to make a very tough, glossy surface, resistant to scratching and quite beautiful. Salt glazes often have an "orange peel" texture to them.

Potters in England, especially around the pottery center of Staffordshire, coated their dishes with a light-colored slip before decorating with other darker slips and over glazing with a clear lead glaze. The famous Toft family of Staffordshire developed a much-imitated method of fancy slip trailing. To create the lively cross-hatches, birds, people, coats of arms, biblical scenes, names, and dates that danced over the surfaces of their cream-colored dishes, they filled quills with dark slip and created the designs with raised dots and lines of clay. Many of these dishes were "too good" or too expensive to use on the table and were purchased to commemorate weddings and births and other important occasions and were hung on the wall.

English potteries were also influenced by the developments on the Continent, and took up both tin glazing and salt glazing. In fact, Cornish mines supplied the tin for Dutch delftware.

And, just as it was on the Continent, Chinese porcelain was all the rage in England. And like potters elsewhere in Europe, English potters felt challenged and threatened. Their customers were entranced with the delicate blue and white Ming wares. The thriving Ming factories were ready to accommodate the tastes and needs of English and European patrons. Chinese pottery decorations became somewhat westernized. Chinese potters added handles to their teacups to satisfy the new European tea drinkers who, not used to hot drinks, preferred to protect their fingers from the heat.

The secrets of the manufacture of porcelain eluded the Staffordshire potters. Nevertheless, by the late seventeenth century, the Elers brothers, John, Philip, and David, had begun to make unglazed stoneware teapots fashioned after the famous Yixing wares. They used a fine, levigated red clay, which, after throwing on the wheel, they shaved thin, and decorated with Chinese motifs such as plum blossoms.

The Staffordshire potteries continued to grow, despite the competition from China. Small family operations became sprawling factories, employing large numbers of workers, often entire families, including the children. Improvements and innovations were made. Calcined flint (this means the flint itself was fired before use) was added to the light-colored stoneware used in salt glazing, which resulted in ware that looked very much like porcelain. Plaster molds were invented, thus enabling the production of very thin-walled pots in intricate shapes that were not necessarily round. Josiah Wedgwood popularized creamware, a light-colored earthenware glazed with a clear glaze that eventually replaced most tin glazed pottery in England and Europe. It was easier to make and more durable. And finally, transfer printing was introduced. This allowed the same design to be mass-produced rather than laboriously hand painted.

At the height of pottery production in Stoke-on-Trent the air was heavy with smoke from the bottle kilns.

The American colonies relied on imports from England, Europe, and China for their tableware. When imports weren't available, or were too expensive, local potters, who brought with them the traditions and skills of their countries of origin, dug clay from nearby pits and made cups, bowls, and dishes for the American market.

The Chinese still guarded the secret of porcelain, even setting up false factories to confuse European spies. The rest of the world had to be satisfied with the tin glaze imitations produced by their own potteries or imports of the real thing carried across the sea in the hulls of sailing ships. Then Johann Friedrich Böttger (1682–1719), a bold young German, stepped in.

Böttger came from a family of goldsmiths and mint workers. He was a bright youngster, mastering mathematics, Latin, and geometry at an early age. He was particularly fascinated with chemistry. The practical career path for a gifted student of chemistry was as a pharmacist, and so the teenaged Böttger was duly apprenticed to a master. But soon, bored with the prospect of grinding powders for medicinal potions and elixirs, he became smitten with the alchemist's obsession and began conducting extensive, secret experiments in his boss's laboratory. By the time he was nineteen, he was engaged in a headlong quest to find the Arcanum, or philosopher's stone, the alchemic secret of turning base metals into gold. Indeed, he had almost convinced himself that he had the secret within his grasp.

At the time, this was not as far-fetched a notion as it might seem to us now. These were the nascent years of chemistry and experimentation. Even the least educated peasants understood that there were occasions when one thing could be turned into another; flour into bread, and clay into stonelike pottery. Why shouldn't you be able to turn lead, or some other material, into gold?

A bit of a showman, Böttger began giving secret performances of his skills, in which he convinced his audience that they were actually seeing him produce small quantities of the precious metal. Not surprisingly, despite his guests' promises to tell no one, word spread.

Soon, even the king had heard rumors of Böttger's talents.

August the Strong, the elector of Saxony and king of Poland (1670–1733) was a man of prodigious appetites. He enjoyed collecting fine art objects, Chinese porcelain, gold, and jewels, indulged in lavish meals, and spared no expense furnishing his estates. Unfortunately, his tastes exceeded his wealth. If a young man could indeed succeed in turning lead into gold, as rumor had it, then he must have him in his employ. The king sent for the alchemist.

Böttger extravagantly pledged that he would produce gold for the ruler within a few weeks. However, weeks and months passed, and Böttger had to resort to one excuse after another, though he apparently did not doubt that, with enough time, he would fulfill his promise. Frustrated, August had Böttger thrown into a dank stone dungeon near the town of Dresden and ordered him to produce gold or face death.

Böttger's incarceration, in several different prisons, lasted for seven years and was at times filled with extreme privation. However, during most of it, he was able to conduct experiments and called upon August to supply him with the necessary materials and assistants. Fearful for his life, yet determined, Böttger set about doing what he had untruthfully claimed to have done all along. An older prisoner, who was a mathematician and alchemist, Ehrenfried Walter von Tschirnhaus (1651–1708), befriended Böttger and became his mentor. Tschirnhaus was interested in porcelain and had spent years attempting to reproduce it. He had extensive notes and knowledge, and he convinced the younger alchemist to take up this pursuit rather than the pursuit of gold.

Böttger suffered bouts of alcoholism, depression, and in time, the loss of Tschirnhaus. Nevertheless, from his work with Tschirnhaus's work, he knew that kaolin was a key component of porcelain. Although he was not able to convince August to free him, he was generally able to convince him to supply his laboratory needs and to give him more time to discover porcelain, which was even more prized than gold, before taking drastic measures against him. Böttger produced a smooth, glassy red stoneware, which pleased the king, but porcelain continued to elude him.

The breakthrough came on January 15, 1708. His notes record these tests:

> N 1 clay only
> N 2 clay and alabaster in the ratio of 4:1
> N 3 clay and alabaster in the ratio of 5:1
> N 4 clay and alabaster in the ratio of 6:1
> N 5 clay and alabaster in the ratio of 8:1
> N 6 clay and alabaster in the ratio of 8:1
> N 7 clay and alabaster in the ratio of 9:1[18]

Böttger goes on to note that the last three were "album et pellucidatum" (he kept his notes in a combination of German and Latin): white and translucent. He had succeeded in producing tiny tiles of porcelain! He was twenty-six years old.

His work, though, was not yet done. He continued to experiment; he needed to refine his porcelain to a reliable, workable body, and to bring his kilns to high enough temperatures. There are contemporaneous accounts of his struggles to bring his kiln up, stoking for days, nearly burning his workshop down.

His achievements were truly wonderful. He was the first European to produce porcelain, and, under August, the first European porcelain factory opened at Meissen. He was disappointed in himself for not making gold and sarcastically inscribed "God the Creator has made a potter from a gold-maker" above his laboratory door. August, afraid that the recipe would get out, kept Böttger and the workers locked up to protect the secret.

In 1724, after Böttger's death, the Meissen factory substituted feldspar and quartz for the less stable alabaster. By now, the factory was producing large quantities of expensive porcelain tableware and figurines. In 1741, after four years of labor, the factory completed one of the largest table services ever produced, the famous 2,200-piece Swan Service for Count Brühl, the factory director. Elegant raised white swans, storks, bulrushes, and mermaids swirled around the water-themed plates, tureens, bowls, platters, teacups, teapots, and other trappings of fine dining. Colored bouquets and Brühl's coat of arms also festooned the overdecorated and extensive table service.

During the last quarter of the eighteenth century, lavish dinner services became a status symbol. Catherine the Great of Russia commissioned Josiah Wedgwood to make her a service for fifty in his creamware. In all there were 952 pieces, with hand painted depictions of the English countryside and frogs on the borders. It was aptly called the Frog Service.

Frederick the Great of Prussia invaded Saxony in 1753. His forces occupied the Meissen factory during the ensuing Seven Years War and halted production. He helped himself to finished cups, plates, and other fine Meissen porcelain. Before his arrival, the Saxons had fled with their special formula clay and notes to protect the secret. The workers who fled, however, settled throughout Europe and brought the secrets of porcelain manufacture with them. At war's end, when the Meissen factory reopened, it was no longer the exclusive European manufacturer of porcelain.

Experiments and innovations continued. Around the turn of the century, bone ash from deceased cows and oxen was plentiful in England, and Josiah Spode added this to his porcelain formula to give it stability. This "bone china," or spode ware, resisted slumping in the kiln and fired white. It remains popular today.

Josiah Wedgwood (1730–1795), whose genius lay in his organizational skills as much as his ceramic skills, created an array of interesting and innovative pieces: functional table settings in cream-colored earthenware (later called the Queen's ware after Queen Charlotte ordered tea and coffee sets), red stoneware, Egyptian-inspired black stoneware, and the legendary jasperware in sprigged neoclassical designs. He pioneered the assembly line in his potteries, separating each of the tasks for more efficiency.

He was so profitable he was able to fund much of his cousin's research.

His cousin was Charles Darwin.

■

SINCE THE MIDDLE of the nineteenth century, factory-made dinnerware has dominated the dinner tables of the industrialized world.

Modern factories are equipped with enormous computer-controlled tunnel kilns, multipart plaster molds, jolleys, extruders, and other machines, much of it heavy-duty steel versions of machines from the past. The majority of the pottery factories bisque fire their dishes at a high temperature, 2,300 ° F (1,260 ° C) or above and glaze at lower temperatures. This allows for brighter colors and for the underside of the dishes to be glazed. The Hall China Company in Ohio pioneered the industrial use of single-firing during the early years of the twentieth century and has been successfully using the process ever since. Their wares are glazed raw and fired to 2,400 ° F (1,315 ° C).

During the 1930's two English women greatly influenced tableware fashion, Clarice Cliff (1899–1972) and Susie Cooper (1902– 1995). Cliff, who was born in Stoke-on-Trent and began work in a local pottery at the age of thirteen, launched her boldly colorful Bizarre ware in 1927. Her work was a radical departure from the delicate floral sprays and romanticized scenes that were prevalent at the time. Cliff used bright primary colors, especially reds and yellows in bold geometric strokes, a reflection of the Art Deco jazz age in which she lived. Her work was widely influential and gave permission to designers elsewhere to break with tradition. Susie Cooper, who was born in Staffordshire, was an equally influential designer. She began her work as a painter in a factory but left to form her own business in which she and a team of forty women painted designs on fired blanks. Soon she began designing her own pots to decorate.

Now, in the early twenty-first century, many couples set up housekeeping without the formal dinner service for twelve that their grandparents thought necessary, yet ceramic plates and bowls and cups or mugs continue to be found in most households. Today's tableware can take on the style of past eras or reflect the new. You might choose china from one of the old European factories—Royal Doulton, Minton, Wedgwood—or inexpensive china made in Taiwan or Japan. You can set your table with brightly colored earthenware from Mexico or Italy, or factory-made dishes from almost anywhere. You might commission a handmade set from a local potter or collect handmade dishes at craft shows. You could even assemble a collection of antique dishes or perhaps settle for reproductions of Thomas Minton's Chinese-inspired late-eighteenth-century Blue Willow pattern complete with faux legend. Our choices reflect our sense of beauty, our need for practicality, our budget, and perhaps our desire for status. Often the dishes we choose reflect our values or sense of ourselves. We can choose from the whole of tableware history for our own tables. Often the "pattern," the decoration, is more important to us than the form.

Hotels and restaurants rely on simple round white plates, "hotel china." "Round and white has been the industry standard for a number of reasons," Julie Gustafson wrote in the online magazine *Hotel & Motel*. "White goes with everything, so there's no need to change dinnerware with the change of carpet and curtains. And chefs love a plain white background to set off their food. Plus it's functional. Round plates fit easily in a lowerator—a cabinet that holds dinnerware. In addition, plate covers, which are ubiquitous in the banquet arena, most commonly are round."[18]

Plain white hotel china has also entered the home, following the earlier popularity of slightly off-white "ironstone" of colonial days. It is dishwasher safe, inexpensive, and unobtrusive.

It wasn't until the seventeenth century that individual plates (from the Latin *platus*) became a dining must in the West. Prior to that, low bowls, or messes, shared with fellow diners were used.

Now, restaurants often have place-keeper plates, called chargers, sometimes made of copper or brass, always oversized, set out on the table for show. Once the meal has begun, the charger is removed, or the dinner plate is set on top of it, and then, during the meal, the waitperson successively removes the used dishes and serves each of the following courses on a fresh, clean plate.

Salt, once served in large salt pigs, moved to individual cellars, then to salt shakers. Today, individual cellars are enjoying a revival.

Place settings for formal occasions still consist of dinner plates, salad plates, soup bowls, cups and saucers, and dessert plates plus a variety of serving pieces such as soup tureen, sugar bowl and creamer, gravy boat, meat platter, vegetable dishes, and perhaps a set of cruets or small salt cellars. Casual family meals often require only a plate and mug, and possibly a small multipurpose bowl for salad or soup or even ice cream depending upon the menu of the day.

During the 1950s and '60s, modern homemakers were tempted by the promise of plastics for their tables, swayed by the claims of advertisers that it was superior because if a child dropped a plate, it wouldn't break, it was dishwasher safe, and it could be taken outdoors for the Saturday night barbecue. Plastic was oh so futuristic.

Alas, in time plastic absorbs odors, becomes dull, and scratches.

They are no match for dishes made of fired mud. Plastic dishes might not break, but neither do they last and they never look as good as the real thing, ceramics. Given the choice, you would no sooner use plastic as paper plates to set the table for a dinner party. For holiday feasts you bring out your best china, lovely blue and white dinner and salad plates, delicate cups and saucers, with matching serving platters and bowls, and set them on the table. You arrange a pale green vase filled with fresh-cut flowers and step back to admire your preparations before your guests arrive. And in doing this you are like every good host or hostess throughout the years, be they kings and queens, laborers, merchants, or peasants living off the land, you serve the best food you can on your prettiest tableware.

A WORD OR TWO

The Invention of Writing and Books

Librarians tucked clay tablets in reed baskets and placed them in the temple storeroom, they stored papyrus rolls in wooden trunks and pots.

—JOHN MAXWELL HAMILTON,
Casanova Was a Booklover

THE ROLE THAT clay has played in the invention of writing and in the proliferation and sustenance of the literary arts begins with the crucial invention of cuneiform by the Sumerians nearly six thousand years ago, and climaxes in the round-the-clock mills that provide the paper for books, magazines, and newspapers today. In the intervening years, clay has played an interesting, though sporadic—and sometimes startling—role.

By the time of the early Tang dynasty, the Chinese were printing books on paper (which they had invented hundreds of years prior) using carved woodblocks. During the Song dynasty, woodblock printing reached its golden age, with highly skilled printers producing numerous books under the aegis of the imperial court. Multiple copies of books of poems, Confucian and Buddhist texts, government records, even playing cards were made available. But printing blocks require ample storage space when not in use, are only adequate for two pages, and of course, can only be used for the book for which they were designed. Between 1041 and 1048 C.E., Bi Sheng (Pi Sheng), a nimble-minded man with no claims to status or wealth decided to address these cumbersome problems. His answer was the remarkable invention of the world's first movable type. Bi Sheng made his invention out of fired clay *four centuries* before Johann Gutenberg made his movable type in medieval Germany!

Shen Kuo, who was a contemporary of Bi Sheng's and probably a friend, described this remarkable achievement in *Dreampool Essays*, a book that Bi then printed using the very process that Shen described. That's right—Bi Sheng invented the world's first movable type, his friend Shen Kuo wrote about the invention, and then Bi Sheng used his invention to publish the book that Shen Kuo wrote. "Pi Shêng," Shen begins, "a man in cotton cloth [i.e., a man of the common people], made also movable type. His method was as follows: He took sticky clay and cut in it characters as thin as the edge of a cash [a coin]. Each character formed as it were a single type. He baked them in the fire to make them hard. He had previously prepared an iron plate and he had covered this plate with a mixture of pine resin, wax, and paper ashes. When he wished to print, he took an iron frame and set it on the iron plate. In this he placed the type, set close together. When the frame was full, the whole made one solid block of type. He then placed it near the fire to warm it. When the paste (at the back) was slightly melted, he took a perfectly smooth board and rubbed it over the surface, so that the block of type became as even as a whetstone."

If one were to print only two or three copies, this method would be neither convenient nor quick. But for printing hundreds or thousands of copies, it was marvelously (literally "divinely") quick. As a rule Bi Sheng kept two forms going; while the impression was being made from the one form, the type were being put in place on the other. When the printing of the one form was finished, the other was all ready. In this way, two forms alternated, and the printing was done with great rapidity.

For each character there were several type, and for certain common characters there were twenty or more type each, in order to be prepared for the repetition of characters on the same page. When the characters were not in use, he had them arranged with paper labels, one label for each rhyme, and thus kept them in wooden cases. If any rare characters appeared that had not been prepared in advance, it was cut as needed and baked with [a fire of] straw. In a moment it was finished.

The reason why he did not use wood is because the tissue of wood is sometimes coarse and sometimes fine, and wood also absorbs moisture, so that the form when set up would be uneven. Also the wood would have stuck in the paste and could not readily have been pulled out. So it was better to use burned earthenware. When the printing was finished, the form was again brought near the fire to allow the paste to melt, and then brushed with the hand, so that the type fell of themselves and were not in the least soiled with clay.[1]

Historians disagree upon whether or not Gutenberg had learned of Bi's invention before making his own momentous contribution to the dissemination of books. We do know that the knowledge of woodblock printing had already come to Europe from China, as had the art of papermaking, by the time Gutenberg began work on his Bible. Whether he was inspired by word of Bi's work is unclear; in any event, he did not use clay.

The advantage that Gutenberg had, and the reason that his printing process had more of an impact on his culture than Bi's had on his, is that Gutenberg had an efficient alphabet of twenty-six letters, while Bi had to work with a vocabulary of three thousand characters. Not everyone in China agreed that Bi's method was more efficient, especially for a shop producing only a few titles. Woodblocks and, later, metal blocks continued to be used in China. By the Ming dynasty, talented printers were turning out beautiful books with multicolored pages.

In this thriving book culture, Ming dynasty scholars and writers enjoyed prestige and lives of tranquil contemplation. They employed the fine art of calligraphy, using brushes and ink, to write their poems and treatises. They paid careful attention to their surroundings, their gardens, and their studies and were especially discerning about the tools of their trade. Here, again, clay is important, as they filled their studies with beautiful accessories of translucent porcelain made by the best potters. Hobson writes, "a special class of porcelain on which great skill was lavished at all times was the furniture of the writing table. This included ink pallets and water-droppers for the same, water vessels for dipping the brushes, shallow bowls for washing brushes, cylindrical vases to hold the brushes, screens to protect the paper from splashes while the ink was being rubbed, rest for the brushes often in the form of conventional hills, wrist rests for the writer, boxes to hold vermilion for sealing and other similar articles generally small and dainty and often of fanciful shapes."[2]

Surely surrounded by such "dainty" items as these, Ming poets were never guilty of messy desks! Unless one is writing in calligraphy, however, such prized pieces of porcelain are not really necessary for the writing life. Common pottery inkwells, however, have been serving writers for thousands of years. As long as books were copied by hand, inkwells of some sort were necessary. Meissen and the great factory potteries of England made inkwells in simple bottle forms as well as more whimsical shapes. American colonial potters made plain dark-glazed, salt-glazed, and Bristol-glazed (in which zinc oxide is substituted for lead) inkwells for their patrons. Often they were short and flat topped, with a large opening in the center for filling and a series of small openings around the rim for dipping the pens. Pens could be left upright in the inkwell when not in use.

Ancient Roman booksellers kept trained copyists as slaves. Often a book would be dictated to a roomful of slaves. Each slave would dip his split-nib pen into a pot of ink made of soot and water and begin writing, much as the later nineteenth-century men of letters dipped their quill pens into their ceramic inkwells. Although the Romans did not have the printing capabilities of the Chinese, in this way they were able to quickly and inexpensively produce many copies of a popular author's work and offer it for sale. Available titles were advertised on the fronts of the shops, and the inventory—long scrolls of papyrus or parchment—were stacked inside, one on top of another. Some books were special orders, or what a modern publisher would call "print on demand," and were only produced when a customer ordered a copy.

Ceramic storage jars have been used to store, preserve, or conceal the written word. In Asia they were used to hold sacred texts for the afterlife. Most famously, cylindrical jars were used to hide (and protect) the Dead Sea Scrolls. The story of the discovery of the scrolls in 1947 has been told many times, becoming almost legendary. Muhammad Adh-Dhib, a young Bedouin goatherd, lost one of his animals in the desolate Judean wilderness between Bethlehem and the Dead Sea. The errant goat had scampered off, toward the steep limestone cliffs that rimmed the edge of the desert. The boy searched and searched and finally noticed an odd-looking hole high up in the craggy limestone precipice. Knowing goats to be curious animals, and good climbers, he tossed a small rock up into the hole. Instead of the bleating of a goat he heard the ping of a stone hitting crockery. He tossed another stone into the hole and again he heard a plink. Curious, he scrambled up the side of the steep rock face and peered into the darkness of the cave. He dropped himself down through the opening.

Once his eyes adjusted to the darkness of the cavern, Adh-Dhib saw that there were tall jars, some with bowl-shaped lids, mostly intact and upright, plus the shattered remains of others. Peering into one of the jars, and thrusting his arm down inside, he discovered what came to be known as the Dead Sea Scrolls. Once word was out, an archaeological furor ensued. Interpretation of the scrolls and possession of them became an international controversy.

More scrolls were found in eleven nearby caves, and in the excavated remains of a settlement nearby. The jars were of a shape unique to the caves and the excavation area, leading archaeologists to conclude that the inhabitants of the settlement were the same people who hid the scrolls. Archaeologists believe the settlement was inhabited from the second half of the second century B.C.E. until 68 C.E. when the Roman incursion began. Most scholars believe that the Essenes, a monastic Jewish sect, lived, studied, and worshipped there.

Edmund Wilson, the American literary critic, visited Qumrân, the site of the excavation, in the 1950s, and was awestruck. "A room with tables and benches constructed of plaster and brick was evidently a *scriptorium*, where the scrolls were copied out. Three inkwells were also found here—one of bronze, which has turned green, and two of terra cotta, turned black—in which there is still some dried ink . . . There is a pottery, with a kind of round nest of stones, which may have held the potter's wheel . . . There was a jar which resembles exactly the jars in which the first lot of scrolls were preserved."[3]

The scrolls, of which there are literally tens of thousands of fragments, include at least parts of all the books of the Hebrew Bible except the Book of Esther and various ordinances, apocalyptic visions, and commentaries. There are multiple copies of some works and mere scraps of others. Whether the scrolls were all copies that were made by scribes or some were collected from Jerusalem and elsewhere is unclear, but it is clear, from the desks and inkwells, that at least some copying went on in the settlement.

It is also clear the jars were used to protect the vast store of literary treasures.

Inkwells served scribes and writers until the fountain pen came into use at the end of the nineteenth century. With the fountain pen, ink could be stored within the instrument, eliminating the need for dipping. This innovation was followed by the invention of cartridge pens and ballpoint pens, rendering ink bottles obsolete and inkwells a collector's item. The most recent enhancement to the ballpoint pen is the ceramic roller ball pen manufactured by several international companies. One manufacturer's online ad exclaims that the pens "are made of precision-formed hard ceramic that will never corrode, flatten or skip—giving you a smooth, clean, fountain pen–like stroke every time. Unlike ordinary metal-tipped pens, which can bend microscopically, Ceramic Tip Pens provide consistent, even ink flow, without dragging or skipping."[4] In fact, this is not advertiser's hyperbole; ceramic tip pens *are* superior to the ordinary metal ballpoint.

But it is the common pencil, the humble yellow implement of grade school days, that most directly uses clay to write. Both the Greeks and Romans used lead disks as markers, the Romans to draw rules on their papyrus to act as a guide so that their prose would march straight across the page. Graphite (from the Greek "to write"), a pure form of carbon, was used to mark sheep in England in the early sixteenth century, a very messy affair for the shepherd, who also ended getting "marked." The earliest pencils encased this graphite in a hunk of wood in order to protect the user's hands and by 1662 a small pencil shop was operating in Germany. A century later, in 1765, another German, Kasper Faber, developed a method of binding graphite with clay to make it more stable and opened a factory to manufacture his pencils.

In 1795 a Frenchman, Nicolas Conte, took this a step further and fired his mixture of clay and graphite in a kiln. He too encased his lead in wood.

An Austrian potter, Joseph Hardtmuth, claimed to have invented the pencil around the same time. He discovered that when he fired his graphite clay mixture at different temperatures, the resulting "lead" came out different hardnesses. This turned out to be much appreciated by users; with soft leads best for sketching and hard leads best for fine drafting. Today we refer to this hardness by number such as the popular No. 2 pencils.

Although we live in an era of cyberspace, e-mail, and rapidly evolving electronic media, the humble pencil, a kiln-fired mixture of clay and graphite encased in an octagonal tube of wood, with an eraser at one end, remains in widespread use. Pencils are inexpensive, safe, and do not leak. They have served us well for hundreds of years. Henry David Thoreau and his dad operated a pencil business in Concord, called John Thoreau and Son, and even won an award for their pencils. Most of us first learned to write our names using a pencil. Some scholars believe that Shakespeare used an early pencil to write his plays.

The oldest known form of writing is the cuneiform of the Sumerians, which was developed during the fourth and fifth millennia B.C.E. in Mesopotamia and used for three thousand years. Writing is a way to communicate; it can convey information from one person to another over time and distances, it preserves memories, knowledge, and records. Before the advent of writing, when many people were apparently able to memorize great epics in their entirety plus the genealogical details of history and directions for everything from how to prepare a holiday feast to when to plant grain far more easily than we can imagine now, there was still an urge to communicate and preserve and surely a desire for aids to memory. We have to wonder if the powerful wall paintings of the Lascaux, drawn in earth pigments, were the recording of events that had taken place, or charms or wishes for the future. Either way, though drawn twenty thousand years ago, they communicate a story to us across time; seeing the stick figure of the hunter lying dead next to the large animal beside him, weapons nearby, we know that both are dead, even though we can no longer discern whether this was depicting an event that happened or an attempt at forestalling it.

Five thousand years before cuneiform tablets were used for accounting records, clay was used to record transactions in Mesopotamia. Because clay was plentiful and easily worked, it could be easily shaped into tokens to represent an item of trade or payment, such as a sheep or fish or cow. If the transaction consisted of ten fish, then ten fish tokens could be fashioned of clay. This, of course, was a bit cumbersome, especially when there were multiple transactions to keep track of. One way to handle this was to then enclose the tokens, in this case the ten fish, inside a clay envelope. In time, the outside of the envelope was marked with impressions of the fish before sealing them inside the envelope so that the keeper of the accounts would know what was inside.

The Sumerians, who lived in city-states and were taxed or tithed by the temple of their city, eliminated the tokens and began to rely on a pictogram of the item being recorded; for instance, the outline of a cow's head would be inscribed in the damp clay. If three sheep were given in payment, then three sheep would be inscribed. But this incised curvilinear writing—though easier to use than tokens—had some drawbacks. The process of incising tended to create burrs in the clay, which were unpleasant to touch and could make reading difficult. And incising was slow work.

Impressing, however, would solve both of these problems.

An impressed symbol was clean edged and quick to make. When the Sumerians began to use a stylus pressed into the clay tablets rather than pulled through it, the pictograms evolved into a series of symbols that looked like horns or bugles, lines with triangles at the ends in a variety of arrangements. The word *cuneiform* refers to these wedge shapes. Cuneiform influenced Egyptian hieroglyphs but with an abundance of papyrus reeds to make sheets of writing material, Egyptians had no need to stray from the cursive lines of their pictograms. However, other peoples, who did not speak the same language as the Sumerians, recognized the value of cuneiform and adopted it for their own languages, including the various Semitic tongues of the Akkadians, Assyrians, and Babylonians. It was also used by the Hittites, Elamites, and others. It was used to write Old Persian.

At first haphazardly arranged on clay tablets, cuneiform was later organized into compartments, which were drawn with lines on the tablet. Eventually, it became customary to read and write horizontally from left to right, top to bottom. Early tablets were small enough to hold in the palm of the hand, say the left palm, while writing with the right, but later they were large enough to require laying flat on a table. Rectangular tablets were most often used, but there are occurrences of cones and cylinders. Students used reusable, round, almost bowl-like clay tablets to copy their teacher's samples. Tablets might be baked or sun-dried. Even sun-dried tablets have lasted in readable condition for five thousand years, though an archaeologist working in the field might sometimes lightly fire them to preserve them.

During the three thousand years that cuneiform was used, it gradually became more abstract. Instead of symbols for three sheep, there might now be one sheep and a symbol to denote three. This led to the development of numbers. A symbol might stand for something with a similar sound, or it might be a syllable or suffix or prefix. It might mean an act, such as eating. Eventually, abstract symbols for syllables developed.

Cuneiform never developed into a *true* alphabet, but it was a robust and flexible writing system. During the first millennium it was used to keep temple records and business accounts. Later it was used for medical "books" and "cookbooks," for official government letters and personal letters, for laws and rules, poems, songs, and epics. The development of cuneiform encouraged the development of mathematics, science, calendars, and literature. Texts were collected, studied, kept in libraries, copied.

Series of clay tablets with wedge-shaped cuneiform writing.

An Assyrian, Ashurbanipal (669–627 B.C.E.), collected thousands of tablets, which he housed in Ninevah in a vast library organized by subject matter. His collection included works in Sumerian and Assyrian, treatises and reports from spies, legal documents, training manuals, poems, myths, even a catalog of the library itself. Though this was Ashurbanipal's library, which he worked on himself, it was, in the true sense of a library, meant for the use of others as well. But, he warned, "May all the gods curse anyone who breaks, defaces, or removes this tablet with a curse that cannot be relieved, terrible and merciless as long as he lives, may they let his name, his seed be carried off from the land, and may they put his flesh in a dog's mouth."[5] This may not be quite what the modern librarian wishes upon the borrower who neglects to return a book or damages it, but I suspect that it is not too different.

Today, the thousands of cuneiform tablets that remain, including many of Ashurbanipal's, offer us an intimate look at daily, municipal, and religious life two to five thousand years ago. We can learn so much more from written records than from artifacts.

The Yale Babylonian Collection holds three Assyrian clay tablets from the time of Hammurabi (eighteenth century B.C.E.), which give thirty-five recipes, the world's oldest known cookbook. Professor William W. Hallo, curator emeritus of the collection, asked French Assyriologist and gourmet chef Jean Bottéro to translate the recipes. Hallo shares the highly seasoned Leftover Stew recipe: "There must also be the meat from a leg of lamb. Prepare the water. Add fat, . . . [break in the text], vinegar, beer, onions, spiney [an herb], coriander, samidu [a spice plant or vegetable], cumin and beetroot. Then crush garlic and leeks and add them (break in text). Let the whole cook into a stew, onto which you sprinkle coriander and shuutinnu [probably a kind of onion]."[6]

For lighter fare, flavored with mint, which to this day is typically added to meals in the Near and Middle East, a cook could turn to the section on Kippu Stew: "If you want to cook kippu in a stew, then prepare them as you would agarukku (a fowl). First, clean them and rinse them in cold water and place them in an open pot. Then put the pot back on the flame and add some cold water to it and flavor it with vinegar. Next, crush together mint and salt and rub the kippu with the mixture. After this, strain the liquid in the kettle and add mint to this sauce. Place the kippu back into it. Finally, add a bit more cold water and turn the entire mixture into the pot. To be presented and then dished out."[7]

Even more extraordinary than the recipes and ledgers that were written on tablets, and that have survived, is the literature: poems, epics, and creation myths, some of which influenced the literature of later ages. The *Epic of Gilgamesh* is best known. It was found in various slightly different versions or with different sections intact, including one version in stone. It is so powerfully written that, even in translation, it resonates with the modern reader. It speaks to us of ego, of friendship, of death and mortality. Remarkably, it includes the modern conceit of a story within a story, that of the Great Flood, and, perhaps springing from an earlier oral tradition, it uses the device of repetition in a way that stirs even the sophisticated twenty-first-century reader.

The *Seven Tablets of Creation*, or the "Babylonian Genesis," begins, "Before heaven and earth had been named, the world was a chaotic mixture of sweet and salty waters personified by Apsu and Tiamat, the father and mother of all gods."[8] The tablets date from the first millennium B.C.E. but archaeologists place the myth at the beginning of the previous millennium. The *Seven Tablets of Creation* and the *Epic of Gilgamesh* demonstrate the rich possibilities and promise that cuneiform writing, wedge-shaped symbols pressed into mud, offered. A line such as, "the world was a chaotic mixture of sweet and salty waters," speaks so much more than a taxman's tally of tokens. Literature became richer, more complex, and more discursive.

Other people developed forms of writing. In Central America, the Maya invented a system of hieroglyphs in the first millennium before the Common Era that included symbols for whole words as well as syllables and was so well organized a system that it could be read aloud by any literate person. By the first millennium of the Common Era writing was highly sophisticated and valued. Even after the height of the Maya civilization had passed, the Maya descendants treasured their books. But when the Spanish conquistadors

entered Central America, they considered the Maya's vast collections—codices written on paper made from bark—sacrilegious and
burned them, to the great horror of the Maya; only a few codices
remain. Luckily, the Maya didn't just write on bark paper. They also
wrote on pots and later carved texts onto stone. We know from the
many portrayals of books and reading on Mayan vessels just how
important writing was to them. Maya pots also contain inscriptions,
sometimes dates, the name of the person in the illustration, and
other references. The people on some pots contain squiggly lines
emanating from their mouths, indicating that they are "speaking" the
text. There are even pots with faux writing, perhaps decorated by illit-
erate potters. The hieroglyphs were a natural for pottery decoration
because they were done with brushes in the same way that the illustrations that wrapped around the vessels were created. Some artists
even got fancy and embroidered their glyphs with extraneous lines.

Fortunately, the Spaniards did not smash all the pots with text,
even heretical texts, and so, particularly in recent decades, using the
pots and stone monuments, scientists have learned to read the hieroglyphs and gain insights into the ancient culture that only contemporaneous accounts can give. The conquering Spaniards were
besotted with the gold of the New World, but in their religious
zealotry they destroyed one of the most important literary archives
the world has known. Without the texts on clay vessels, and stone
monuments, Maya literary culture would be lost forever.

In the years since the beginning of the Common Era when
cuneiform was written on clay tablets for the last time and today
when paper is the medium, clay has not been the surface of choice
for written materials. Yet, as it turns out, clay is a key ingredient of
modern paper.

"Most paper," writes Victor Greenhut, who holds the Corning/Saint-
Gobain–Malcolm G. McLaren Distinguished Chair in Ceramic Engineering and serves as the executive officer of the Department of
Ceramic and Materials Engineering at Rutgers University, "has 2%–10%
ceramic as a so called filler material; some papers have even more. The
ceramic is actually a very important active ingredient providing paper
with opacity and whiteness, while controlling the flow of ink in

writing and printing. Without the ceramic, ink would be absorbed by and smear into the paper. The ceramic also can provide color. Ceramic has been used for many centuries as an important paper additive, with china clays, such as kaolin, having very long histories as the applied material."[9]

He goes on to explain that titanium dioxide, which is also used as a whitener in toothpaste and paint, replaced kaolin as a whitener in paper during the middle of the twentieth century. The clay industries suffered economic losses as the use of kaolin declined. However, Dr. Greenhut and others at Rutgers discovered that by calcining (heating) kaolin under tightly controlled circumstances with the addition of water vapor and mineralizers, kaolin is actually superior to titania and produces a better grade of paper. "After about a decade, the resulting calcined kaolin powders have virtually displaced titania as the paper additive of choice for whitening, brightening, and opacification."[10]

Bentonite, a fine-particled clay (montmorillon) that stems from decayed volcanic ash, is also important to the paper industry. Because it is highly absorbent it is very effective in controlling the resins in wood pulp from gumming up the machinery during the paper-making process. It is crucial to recycling and is used to de-ink printed paper. And bentonite is a component of self-copying paper such as used in receipts, invoices, and cash register tapes.

Clay is also used to coat paper. High-grade paper, like that used in art books or for a thesis or resume, is given a finishing layer containing clay before the paper is cut to size. This makes the paper smooth and satiny.

In addition to being a key component of paper itself, ceramic materials are used in machinery throughout modern paper mills because they wear better than other materials such as metals and are resistant to corrosion. The chipper blades, which can render a log the size of a man into chips in a matter of a few seconds, are made of a durable ceramic-metal composite. The huge vats for mixing the pulp are usually lined with bricks, to resist corrosion. The slats and plates of the papermaking machines themselves are made of ceramic.[11] We have come a long way since the time when a Sumerian farmer first recorded the number of goats he sold his neighbor by making marks in a lump of damp river mud. Yet, the paper that this book is printed on, the ream of paper you keep near your computer, are not really as different from the ancient tablets as we might at first think, for they too are made of clay. However, today's paper readily decomposes. It burns into a pile of ash. One of the texts consulted for this chapter, *Voices from the Clay*, published in 1965 by the University of Oklahoma Press, proudly states, "The paper on which this book is printed bears the University of Oklahoma Press watermark and has an effective life of at least three hundred years." Many of yesterday's cuneiform tablets are still readable after *thousands* of years.

Today, a great many baked clay tablets reside in modern climate-controlled museums and archives, awaiting translation. Unfortunately, in 2003 an unknown number of them were destroyed or looted from Iraqi museums. Perhaps some of these lost clay tablets held correspondence between one Babylonian and another, or the household records of a farming family. Or perhaps we have lost a work of literature as extraordinary as the *Epic of Gilgamesh*.

We will never know.

Perhaps Ashurbanipal's curse will come down upon those responsible for the loss.

THE MOST POPULAR
BUILDING MATERIAL
Cities, Walls, and Floors of Mud

The third little pig met a man with a load of bricks, and said:
"Please, man, give me those bricks to build a house with."
So the man gave him the bricks, and he built his house with them.

—JOSEPH JACOBS,
The Story of the Three Little Pigs

I N THE WELL-KNOWN children's story, three impoverished little pigs are sent out into the world by their struggling mom to make their own way. The first little pig meets a man carrying a bundle of straw, and asks for the straw to build a house. The second meets a man carrying a bundle of "furze," or evergreen sticks, and asks for the sticks to build a house. The third meets a man carrying a bundle of bricks, and asks for the bricks to build his house. Each of the men the pigs meet generously hands over the bundle of building material. But, as every child knows, danger lurks nearby in the form of the hungry big bad wolf. Sure enough, the wolf arrives and after demanding entrance, he huffs and puffs and blows the straw house down and eats the first little pig. He then dashes off to the stick house and does the same thing: he huffs and puffs and blows the house down, and eats the second little pig. But when he gets to the sturdy brick house, something different happens. No matter how hard the big bad wolf huffs and puffs, he cannot blow the brick house down. The house made of clay is a fortress against outside threats.

Houses have been made of straw, sticks, skins, bark, stones, fired and unfired bricks, and piles of mud for thousands of years. What material is used is usually dictated by what is most available, and the way the inhabitants live. Hunter-gatherer and nomadic cultures require shelter that can be easily transported or immediately occupied when the hunters or families reach a particular area. Tents are lightweight and can be readily set up almost anywhere. Caves and rock shelters, where geologically prevalent, offer protection from the elements and make good stopping places.

As people began to remain for longer periods or indefinitely in one location, they built more permanent houses, often of mud. Although mud may intuitively seem to be ephemeral, a mud building, with care, can easily last for centuries. Sun-dried clay does indeed dissolve when soaked, but it takes extended exposure to a lot of water to return an adobe brick to mud. A potter can dunk a bone dry pot into a bucket of water and remove it intact; if the pot is left in the bucket, it will slowly disintegrate. The same is true of mud walls, although, because of the thickness, it would take much longer to dissolve.

Conversely, mud walls dry very quickly. If you sprinkle drops of water onto an unfired pot, the spots left by the droplets will vanish almost instantly. Interestingly, this is not true of other "more permanent" materials such as cement or concrete, which won't dissolve, but which also hold the moisture for a longer duration.

Mud walls are sturdy, can support a surprising amount of weight, and, when two or more feet thick, they provide wonderful thermal mass. A mud wall soaks in solar heat during the day and gives off warmth at night. Today, about a third of the world's population—some estimates are as high as one-half—live in houses made of unfired clay. If you count fired bricks, the proportion of people whose houses are made of clay is even higher.

There are several basic types of clay construction.

Wattle-and-daub (also called jacal or bajareque) houses, barns, and stables are made by coating a woven structure of sticks or reeds with mud. Cob houses are constructed of balls or chunks of clay mixed with straw or other organic temper, and a bit of manure or sand, and stacked into walls and smoothed together. Adobe bricks are similarly tempered, shaped into blocks, dried in the sun, and then stacked and mortared together into walls. Rammed earth, or pisé, walls are made by compressing the earth with force, often between forms. And fired or burned brick buildings are made of walls built from bricks that were previously fired in a kiln.

Each of these methods is ancient.

The Vinca were a prehistoric people who lived in snug wattle-and-daub houses in the central Balkans seven thousand years ago. They built their homes by making a frame of timber posts that they then interwove with twigs, making the wattle. The whole structure was plastered with a thick layer of clay, the daub. Smooth plaster floors were laid down over a base of broken pots, stones, wooden beams, or a combination of the three. A house might have one room or several rooms separated by interior wattle-and-daub walls.[1] Excavations in Bulgaria have uncovered whole villages built of wattle and daub dating from the fifth millennium B.C.E. Here again, the houses, shops, and temples were framed in timber and the walls were made of basketry coated on both sides with a layer of clay.

Wattle-and-daub houses require regular maintenance if they are to last. Archaeologists have found some houses with indications of as many as twenty coats of mud, suggesting an annual renewal. This is common throughout the world wherever sun-dried mud houses are found. Often the annual rite of plastering the walls is associated with a particular season, such as the end of the rainy months. Though mud itself is fireproof, the threat of a spark from the oven, located inside the house, igniting the thatched roof was constant. However, because wattle-and-daub houses were built of abundant local materials, and were quick to erect, a family or village who suffered from fire could rebuild right away.

The Navaho built a less complex mud hut called a hogan. The earliest hogans were made by fastening three upright forked poles together at the top, and then laying other poles across, between, and around them. The low structure was covered with mud that was tamped into place. An opening was left for the doorway. Later hogans were built with four forked posts instead of three. First a round pit two feet (sixty centimeters) deep was dug. Then the posts were erected near the edge of the pit, about ten feet (three meters) apart. Next, two parallel poles were laid across the four uprights. Branches were laid across the roof poles. More branches were stuck in the ground around the perimeter of the pit, and woven into the roof branches. An opening facing east was left for a door. Lastly, the structure was covered with moist earth. Eventually, hogans were made of logs covered with mud, domed on the top, with an opening in the roof for smoke to escape, and an entrance door covered with a blanket.[2]

The Hohokam of northern Mexico and the southwestern United States also built wattle-and-daub pit houses. They used the mud that they removed to form the base or pit of the house, to cover the posts and woven brush of the walls and roof. Similar wattle-and-daub pit houses have occurred throughout the world.[3] These shelters are easy to build, efficient, and provide warmth and safety. Some of the earliest colonists to settle in New England spent their initial frigid winter in similar wattle and daub pit houses.

Marcus Vitruvius Pollio, the Roman architect and critic, disparaged wattle and daub in his famously opinionated *Ten Books of Architecture.* "As for 'wattle and daub' " he wrote, "I could wish that it had never been invented. The more it saves in time and gains in space, the greater and the more general is the disaster that it may cause; for it is made to catch fire like torches. It seems better, therefore, to spend on walls of burned brick, and be at expense, than to save with 'wattle and daub,' and be in danger."[4] We are not sure when Vitruvius lived, but most historians believe it was during the first century B.C.E. Despite his advice, and the longevity of his book, wattle-and-daub construction has continued to this day.

Many of the picturesque pubs and flower-bedecked inns of the British Isles, romantic with dark beams and contrasting tan or white panels of plaster, so enticingly photographed for guidebooks, are also of wattle-and-daub construction. In England, wattle and daub is often associated with the Tudor period (1485–1603). In the United States faux wattle-and-daub mansions (usually stucco) are called "Tudor style."

In the authentic wattle-and-daub buildings of the United Kingdom, a heavy post-and-beam frame was crossed with lightweight pieces of wood called withies. Often the lower walls of the house were filled in with stone set on a stone foundation. The wattles were woven of willow or hazel staves, each wattle the size of the opening in the frame between the upright posts or timbers and the withies. Once the wattles were set in place and attached, the builders quickly and simultaneously slathered both sides with daub, a local mixture of clay, straw, hair, and sometimes manure. To prevent cracking, the walls were protected and dried slowly. Once the wall was thoroughly dry and consequently hard, it was coated with plaster or lime.

Wattle-and-daub houses can last for centuries, and indeed, in the British Isles, the Cayman Islands, and elsewhere, there are many historic wattle-and-daub houses that have sheltered generations of families. They are invitingly comfortable and weather tight.

Modern proponents of wattle and daub, such as Ian Pritchett, managing director of IJP Building Conservation in England, point out that wattle and daub is environmentally friendly.[5] The materials are organic and renewable and do not require vast amounts of petrochemicals in their production. Material for wattle can be quickly grown and harvested, especially if one is using willow from a coppiced tree. Clay is all around us. Few trees are needed for the frame, especially compared with a wood house. Simple repairs and maintenance require few tools. If the daub dries and shrinks, one can simply mix up some more and squeeze it into the crevice.

Rose-covered cob cottages also feature heavily in glossy travel guides, inviting tourists to step back in time. In fact, there are about fifty thousand cob buildings extant in England today, most dating from the eighteenth and nineteenth centuries,[6] though the English used cob to build houses (and outbuildings) at least since the fifteenth century. Cob houses do not require a frame; instead the cob was made on-site by digging the clay from the same property where the building was desired. The clay was then mixed up and roughly formed into balls, and the walls were built up from layers of these balls. Often pots, such as pans or jars, were embedded into the cob walls during construction to serve as pantries, cupboards, and display niches once the house was complete.

Heywood Sumner, a Dorset man concerned about the status of cob building in the early 1920s, described the process: "This traditional craft has been dying out. Mud walls should be made of sandy, clayey loam with small stones in it; and with heath, rushes and sedge-grass or straw, thoroughly puddled into the mass by trampling. In the best-made mud walls this was dubbed and bonded by the mud-waller with his trident mud-prong in successive layers on the wall he was building. About two feet, vertical were raised at a time (a 'rearing'), then left for ten days before the next rearing was raised on it. Walls built thus, on heathstone or brick footings, stand well. But often they were raised without any footings, and by inexperienced 'mudders' who used the wrong sort of clay; who did not temper it stiff with heath, and who could not build a wall with a mud-prong, but trusted instead to board 'clamps,' and thus this serviceable walling material has been discredited, most unfairly; mud walls that have been built well stand firm and impervious for generations, and provide warmth in winter and coolness in summer within the cottages which they surround, and they cost less than walls of any other material locally available. There is excellent mud-walling still being raised, and for such requirements, I should go to Verwood, who has inherited the knowledge of his craft, and can point out this and that, and the other mud-walled, thatched cottage in his native village, as built by his grandfather, his father, or himself (to the last he has been recently been adding). An outer rough coating of plaster and pebble-dash fortifies the weatherproof nature of mud-walls, and veils a coarse material with a serviceable finish."[7]

Another Englishman, Sidney Frampton, writing ten years later, gave this set of directions: "First you prepare your foundations, you dig them and put them in concrete, some people used to put sandstones into the foundations but this is not so good because later you tend to get cracks in the walls of the buildings in the line with the joins in the sandstone.

"Then you get a cartload of clay or loam, and tip a cartload onto each way (where you intend the walls to be). Then you use water to make the clay wet and then tread the mixture well. You tread it first, then turn it over with a prong and tread it again. When you've done that a time or two you put green heather into it. Heather is used because it never rots. Then you turn it again and keep on treading it and adding more heather until the texture is right. That's until it's stiff enough to stay on the prong. Now you can start. You start by plumping it down on the wall, eighteen inches high all along. You must do this in one day. Then do another wall in another day, and so on until you have four walls at the end of the fourth day. On the fifth day, if it is dry enough, you can add another eighteen inches to the top of the first wall.

"Of course, this is a job for the summer-time, you must have dry weather. So you can see that this is a long hard job. But for warmth, you can't beat a cob wall."[8]

Though skill was required to build a cob house, barn, or workshop, it was something that a reasonably handy farmer could do himself with the help of a few friends or his family. But fancy homes were also built of cob. Sir Walter Raleigh had such fond memories of the cob manor house where he grew up, that in around 1600 he offered to buy it from its then owner, for "whatsoever in your conscience you shall deeme it worth." He was out of luck, however, as the current resident had no intention of moving out of his lovely upscale mud home. The house, called Manor House at Hayes Barton, is still in excellent condition today.[9]

New cob construction declined by the end of the nineteenth century and ceased by the middle of the twentieth century. Then, in the seventies, as "alternative, earth friendly" building methods were rediscovered in the West, interest in cob was revived. Today, there are workshops and several how-to books available. A quiet cob renaissance seems to be taking place, particularly in Oregon, but also in England, Australia, and New Zealand. However, many American building officials remain skeptical, and sadly uninformed, about mud as a building material.

In India, China, Pakistan, Afghanistan, Yemen, the Near and Middle East, Africa, and parts of South America the tradition of building with sun-dried mud has been unbroken for millennia. In Yemen, buildings made of mud balls can be thirteen stories high. The Taos Pueblo, the oldest continuously inhabited community in North America, built between 1000 and 1450 C.E., has mud walls several feet thick. The pueblo consists of numerous connected houses and is five stories high. There are no doors or openings between the houses, and originally a home could only be entered through the roof, which was made of heavy beams crossed with thinner aspen poles packed and layered with mud.

The British archaeologist Kathleen Kenyon discovered ten-thousand-year-old sun-baked bricks in the city of Jericho. This ancient oasis city of biblical fame lay 820 feet (250 meters) below sea level, near the River Jordan and the Dead Sea. It was an island of lush green in the arid desert, a crossroads of trade, subject to attack, invasions, fire and earthquakes. The earliest adobe bricks found in Jericho, dating from about 8300 to 7600 B.C.E., were made by kneading and patting soft clay into loaf shapes of roughly the same size. The bricks were set out to dry in the hot sun before laying them up in a wall mortared with more clay. They were used to build somewhat circular houses, tall towers, and wide protective walls. Remarkably, these sun-baked bricks predated any notion in Jericho of using clay to make pottery.

Later bricks found at Jericho, dating from about 7600 to 6600 B.C.E. were also hand formed, but more standardized in size, and longer and thinner. They still bear V-shaped thumb marks from the brick makers.

Çatal Hüyük, the vast mud brick city discovered in Anatolia, dates, in it deepest layer, from about 6385 B.C.E. This Neolithic metropolis consisted of houses, workshops, stores, and, it is believed, temples, all built of sun-baked bricks. Like Taos Pueblo, the buildings were adjoined. There were no streets or alleys or walkways. There were also no doors. Instead, each house had an oven along the southernmost wall. The smoke outlet overhead in the roof also served as the entryway, with a ladder down into the house. All the houses at Çatal Hüyük follow nearly identical floor plans.

Like Jericho, the bricks at Çatal Hüyük were hand formed. They were tempered with chopped straw or reeds. The mortar was made of ash and ground animal bones. Houses seemed to have lasted about 125 years, at which point they were knocked down, and rebuilt in exactly the same spot. Over the centuries, as generations built on the foundations and rubble of their predecessors' homes, the mound (or "tell") grew in height.

We do not understand why the inhabitants, who dug their clay from the surrounding countryside, and who enjoyed trade with other cities, kept building in the same spot, one level upon another. This was true in other Near Eastern cities and towns, which also formed mounds or tells.

Çatal Hüyük was inhabited for about eight hundred years. It has attracted speculation, mythologizing, and inquiry because it holds so many mysteries. The archaeologist James Mellaart, who excavated Çatal Hüyük in the late 1950s and early 1960s, uncovered dozens of stunning paintings, some four feet in length, on the white plastered interior walls of the mud houses. Using vivid pigments, prehistoric residents of the town enlivened their rooms with depictions of groups of hunters wielding lassos, fleeing deer, hulking vultures, leopards, enormous bulls, "goddesses," and women's breasts. Some of the buildings contain protruding plaster bulls, often with real bones inside. The houses all contain raised platforms, thought to be beds, and within the platforms human bones were interred, an average of eight bodies per house. It appears the plaster was reopened and closed periodically, perhaps to add more bodies.

Thousands of ceramic figurines have also been recovered from Çatal Hüyük, mostly females with exaggerated buttocks and breasts.

Were the residents creative interior decorators—or were they adherents of a goddess and bull religion? Both Mellaart and Ian Hodder, the lead archaeologist currently working at Çatal Hüyük, believe that the images do have a spiritual connotation and that the inhabitants followed a goddess religion.[10] One doesn't have to disagree with them to also conclude that the Çatal Hüyük householders' home decorations added appeal and vitality to their rooms.

Around 5000 B.C.E. brick-molds came into use in Mesopotamia.

Now a single brick maker could produce hundreds of identical bricks in one day. Molds were made of wood. They were rectangular, open on the top and bottom, and might have had a handle. Working in a cleared area, likely strewn with straw, a brick maker would overfill the mold with soft clay mixed with sand, manure, or chopped straw and then quickly scrape across the mold with a board, pushing the excess clay over the edge. He would then deftly remove the mold, leaving the wet brick to air-dry in the sun. The brick maker would repeat the steps, setting out row upon row of fresh wet bricks to dry. This process spread to Egypt, where stone was the preferred medium for royal buildings but sun-baked brick was used for everything else. Egyptian tomb paintings illustrate the work.

Sun-dried mud bricks are more convenient to carry than wet balls of clay. Their manufacture enabled builders to work distances from their source of clay. Molded bricks save the brick maker time and effort, are simpler to lay than hand-sculpted bricks, and they require less mortar to level. The exact same brick-making method is used throughout the world today, though sometimes a double mold is used so that two bricks can be made at once—more than two are too heavy and unwieldy.

Adobe, or sun-baked, brick buildings come in a dazzling array of styles, shapes, colors, and uses. In the hands of some, adobe is an art form. Extraordinary adobe mosques, some hundreds of years old, serve the Muslim communities of Mali, Niger, Nigeria, Morocco, Afghanistan, Ghana, Iran, and elsewhere. Many have vaulted or domed roofs, geometric designs made by indenting some bricks and protruding others, phallic pinnacles and turrets, niches, arches, stairs, and towers. They stand isolated under the searing sun of the desert and in the midst of bustling cities.

In Pakistan, adobe houses are cooled with wind catchers. Usually, but not always placed on the roof, these are two-sided structures with a slanted top that catches the wind and forces it down into the house. The size of the wind catcher is dictated by whether the house is in a high or low wind area; the larger scoops being built where the breezes are slight, smaller where gales whip across the countryside.

In India mud houses are made of both cob-type balls of clay and

The century-old Grande Mosque in Bobo Dioulasso, lovingly sculpted out of mud.

adobe bricks. Women decorate the insides of their mud houses with lacy raised grids, garlands, swirls, and arabesques of clay, which they coat with a gleaming whitewash and in some areas, decorate with bits of mirrors. They make chests and dressers and benches of adobe. Elsewhere in India, the walls and floors are burnished to a soft sheen by rubbing the adobe much as one would polish a water jug, and exterior adobe steps lead up to flat roofs.

Large, highly decorated palaces and colleges in Nigeria are fashioned of adobe. Raised, swirled motifs surround an imposing college entrance in Katsima. A cavernous ribbed vault welcomes guests in a reception room at the Palace of the Emir in Kano. In Peru, Bolivia, and Mexico, Spanish-influenced adobe intersects with the bold carved-

Lovely mud house with outdoor staircase in India. The rope bed is being used to dry clay while freshly decorated pots dry in the sun.

decoration design influences of pre-Columbian mud buildings such as the twelfth-century temple pyramids of Trujillo, near the arid Pacific coast. The farmers of rural western China build sun-baked mud walls to enclose their compounds, and live in sun-baked mud houses, much as their ancestors thousands of years in the past.

Three thousand massive adobe pigeon towers, many thirty feet in diameter, stood in Isfahan, Iran, at the beginning of the eighteenth century. A typical tower housed a thousand pigeons, each with a personal niche and a cuplike perch that jutted out from the base of the niche. In a year, a pigeon tower could garner three tons of pigeon dung to fertilize the surrounding melon fields and orchards. These enchanting mud towers were engineering marvels and stood until the incursion of chemical fertilizers into Iran.

The world's first windmills, also made of sun-dried brick towers, were built in western Afghanistan near the Iranian border in the ninth century, four hundred years before windmills were built in Europe. The mud windmills, topped with woven reed sails that turned horizontally, harnessed the strong prevailing winds to grind corn and pump water.

Wherever they are found, in whatever time period, unfired mud houses have an organic quality. Edges are rounded. Shelves and benches flow into walls. Despite the right angles inherent in adobe bricks, in the buildings themselves, the angles are softened into gentle curves. Literally and metaphorically of the earth, earth colored, these houses are a part of their surroundings, not apart. Affordable and of low ecological impact, adobe houses do have the drawbacks of being vulnerable to floods and earthquakes. In recent years, whole towns have been tumbled to rubble by quakes in Afghanistan, China, and Iran. Though adobe is vernacular, perhaps a twenty-first-century engineer will find a solution.

Fired bricks, also called burned bricks, are more permanent, more expensive, and require more skill to make and use than unfired bricks. Fired bricks dating from about 5000–4500 B.C.E. for use in a drain were found in Mesopotamia, but no others have been located. However, fired clay cones or cones dipped in bitumen were sometimes sparingly interspersed between adobe brick courses, as ornament or strengthener. It was not until the Uruk/Djemdet Nasr period (3100–2900 B.C.E.) that fired bricks were made in quantity. Burned bricks

An efficient brick-covered drain in Mohenjo-Daro in the Indus Valley in the third millennium B.C.E.

appeared in Mohenjo Daro, the urban center of the Harappan in the Indus Valley in the third millennium B.C.E. We do not know whether they learned to use fired bricks through trade, or they came to the idea independently.

The most spectacular and best-known structures of the Fertile Crescent were the huge stepped pyramids, called ziggurats, presaged by raised temples, such as the Painted Temple at Uqair (C.3200–2000 B.C.E.), which was built on a massive mud brick platform reached by mud stairs. The walls were brightly colored and decorated with mosaics. Drainage ditches protected the mud from floods.

Ziggurats were imposing edifices also built on wide bases. They reached progressively skyward to a top platform where yearly renewal ceremonies took place. Steep staircases ascended to the temple platform. These were arguably the pinnacle of adobe construction and required planning and hard labor. In the Cassite Ziggurat, a layer of reed matting was utilized after every fifth course of bricks, and reeds were used to tie the whole structure together.[11]

It would have been prohibitively expensive both in time and fuel to build a ziggurat entirely of fired bricks, but some, like the Ur-nammu Ziggurat, had a facade or revetment of fired bricks with an inner core of adobe bricks. This added protection and permanence.

James W. P. Campbell describes the enormous undertaking that building a ziggurat entailed in his delightful book, *Brick: A World History*: "It has been estimated that the ziggurat at Babylon contained some 36 million bricks and that, of these about a tenth were fired and the rest were mud-baked, requiring some 7,200 working days to mould the fired bricks and 21,600 working days for the mud ones. By the same calculation it was estimated that the ziggurat would have employed some 1,500 workers (87 moulders, 1,090 masons and 404 porters) just to make and lay the bricks. From various documentary sources we know the names of some of the trades involved: digger of mud, mixer of mud, mud specialist, porter of mud, maker of baskets for transporting mud/bricks/mortar, brickmaker, architect/chief builder."[12]

Later, Babylonian brick makers made stunning glazed bricks with relief designs, which they assembled into murals for the interior palace chambers of the king, Nebuchadnezzar II (604–562 B.C.E.) and placed on the famous Ishtar Gate to the city. The bricks were probably laid out on a flat surface such as the floor of a workshop while still fairly wet. The artist or brick maker would have taken care to assure that the bricks were all the same size, and lay close. Then, probably using a template, the desired images would have been lightly scratched into the clay to serve as a guide, before carving and modeling could commence. Once the relief work was done, the bricks were glazed and separated to dry and then fired in a kiln. Some authorities believe that, instead, large, thick, flat panels of clay were first set on a level surface such as a workshop floor and cut into bricks after the designs were complete. Either way, the technical proficiency—in forming, glazing, firing, and then reassembling and affixing the bricks—was an unprecedented feat.

The walls of the throne room were faced with bricks glazed in teal. A row of raised lions, glazed white, tales switching upward, mouths open as if roaring, marches across the center of the wall. Above and below the lions are rows of white and yellow daisylike flowers and geometric bands of orangey yellow.

Artisans working under the Persian king Darius I (521–486 B.C.E.) perfected glazed bricks further. They made their bricks wedge shaped so that the faces would fit tightly without showing the mortar.

Mortar for fired brick was quicklime, which could be made by firing lime at the same time and temperature that the bricks were fired, or bitumen, which could be found in surface pools in this oil-rich area. The use of fired bricks led to the separation of brick making and brick laying. Bricks were used much as stone, and required the skills of a mason, including cutting fired bricks.

During Nebuchadnezzar's reign, fired bricks cost two to three times as much as adobe bricks. Earlier, during the dynasty of Ur (2111–2003 B.C.E.), fired bricks were thirty times more costly than sun-baked bricks. In Egypt, where sun-baked bricks continued to prevail, burned bricks had little appeal because of the availability of stone and the expense and trouble of firing bricks.[13]

A good roof is one of the most important parts of a building. Old-time English cobbers used to say a house needed a "good hat and good shoes" if it wasn't to fall down. This has always been true. The roof and foundation protect the walls from groundwater, dampness, rain, and snow. Ancient roofs were made of thatch, straw, wood, wattle and daub, logs covered with lighter sticks covered with layers of adobe, logs covered with lighter sticks covered with adobe covered with tar or bitumen, mud brick vaults and domes, and later fired brick vaults and domes.

Terra-cotta roof tiles were a major improvement. Tiles dating from perhaps as early as 2600 B.C.E. have been recovered in the Peloponnesus peninsula in southern Greece. By the golden age of Greece, roof tiles were common. The Greeks made three styles of roof tiles. The simplest was the Lakonian tile, which was made of a rectangle of clay curved into a half cylinder. These were set on the roof, with two rows placed alongside each other, the curves open to the sky, and a third row positioned so that the curve covered the join of the first two rows. This was repeated, two concave rows protected by a convex row. Today, we call this sort of tile a Spanish or half round tile.

A modification of this style of tile roof, called the Corinthian system, utilized two rows of flat tiles with turned-up edges covered by an inverted V-shaped ridge tile. A third system combined both types, using curved tiles for the ridge with the flat tiles with turned-up edges of the Corinthian system. This is called Sicilian.

Today, we associate skylines of red terra-cotta tiles with the scenic hill towns of the Mediterranean, especially Greece, Italy, and Spain, and with Mexico and South America. Though now factory made, they are essentially unchanged from those of ancient Greece, which some romantically imagine were formed over the thighs of goddesses.

A variation on the curved tile, which the ancient Greeks did not adopt, but which was used by the Flemish during the Middle Ages and perhaps earlier in the Orient, is the sideways S-shaped interlocking pantile. More difficult to manufacture than the simple curved tile, the pantile was easier to install and more watertight.

A simpler and later roofing tile is the English, or "plain," tile, which resembles shingles or slates. These are made flat or with only the slightest curve. They might have a nib, so that each succeeding tile could hook over the previous, or a nail hole for affixing the tile to

This romantic French farmhouse is built of cob with a fired brick chimney and tile roof.

the roof. English tiles are handsome but with their weight, they require a sturdy house to support them. They can easily be made by hand, a laborious and tedious process, and so have some appeal for the time-rich and hardworking self-sufficient homesteader.[14]

The Chinese were not content with the mere practicality of burned clay roofs. They wanted something to look at!

By the Neolithic era, many Chinese houses were timber framed with paper, mud, and later burned brick infill. The walls were not load bearing. The roofs, supported by the upright timbers, were framed with a series of four-part brackets. They had wide overhangs and upturned eaves, like an open paperback book facedown on a table. At first sealed with a layer of mud, they later turned to fancy burned tiles, which they enhanced with ornamental disks along the edges, and "coiled dragons in sunken relief."[15] They topped their roofs with "finials and ante-fixal ornaments ... [of] cleverly modeled figures of mythical personages, birds, and animals."[16] And though many of their roof tiles were the charcoal grays of a reducing kiln and matched the color of their bricks, they went further than other cultures and often glazed their roof tiles in gleaming and intense shades of green, yellow, blue, or red.

Burned bricks were being made in China at least as early as the Warring States period (475–221 B.C.E.) and possibly earlier. Until the Western Han period (206–224 B.C.E.) the Chinese built their walls using only one tightly fitted layer. They made hollow as well as solid bricks, and during the Eastern Han period (25–220 C.E.), made innovative tongue-and-groove bricks.[17]

The most impressive brick structure, not only in China, but in the world, is the Great Wall, which is one of the few structures on earth that can be seen clearly from a satellite. It is actually multiple walls built of unfired loess (a fine-grained, loamy clay), bricks, and burned bricks fired in reduction kilns during the reigns of several dynasties. The wall was begun during the seventh and sixth centuries B.C.E. during the Spring, Autumn, and Warring States periods and was made of unfired bricks. The emperor Qin Shihuang (259–210 B.C.E.) linked these early mud brick walls together and extended them to make the First Great Wall, a three-thousand-mile (five-thousand-kilometer) fortification against marauding invaders from the north. The wall was extended and strengthened during the Western Han dynasty (206–24 B.C.E.). Further, more northerly branches were built during the Jin dynasty (1115–1234 C.E.).

During the Ming dynasty (1368–1644 C.E.) the army built what we think of as the Great Wall of China today. They used rocks and burned bricks. The warriors manufactured the bricks themselves, stamped each brick with the date and the army unit, and then fired them in reducing horseshoe kilns. Usually, they built two parallel walls that they filled in with rammed earth, rubble, and stones. Rice gruel was added to the lime mortar in the belief that it added strength (perhaps it was really because the soldiers could not bear to eat it). The wall was wider at the base for strength and varied in height and width depending upon whether it was crossing the ridge of a precipitous mountain or in a more vulnerable valley. The soldiers built guard towers at intervals along the wall. To enter a tower, a guard would climb a wooden ladder to an elevated door opening. When under threat of attack, the guards withdrew the ladder, making the tower safe and impenetrable.

Sections of the wall have deteriorated over time. Thankfully the government has undertaken restoration. Much of the Great Wall, however, still stretches imposingly across the mountainous countryside and many thousands of visitors come from all over the world to see it and marvel at the extraordinary building feat.

The Chinese pagoda is also much admired around the world. These Buddhist towers were derived from the Indian stupa, which symbolized the sacred and mythological Mount Meru. The oldest standing stupas in India date from about 250 B.C.E., little more than two centuries after the death of Siddhartha Guatama (c. 563–c. 483 B.C.E.). The first stupas were made of mud and stone. Later stupas were built of burned brick. They rise skyward, the warm rust color of terra-cotta, with stepped levels, massive exterior brick staircases, and multiple ascending pointed vaults.

In the ancient city of Pagan, the seat of a kingdom that flourished nine hundred years ago in what is now Myanmar, five thousand stupas and temples pierced the sky. Today, about two thousand stupas and temples remain, "forming one of the largest collections of ancient brick monuments in the world."[18]

Both the stupas and temples of Pagan are complex, with hidden passages, "inaccessible spaces and interconnecting vaults."[19] Glazed tiles depicting the five hundred lives of the Buddha graced the exterior walls of some of the stupas. Burned brick stupas of various designs, some with stone bases, others with elephants clustered on a midpoint terrace, appear throughout Buddhist lands of southeast Asia, each reaching heavenward like holy Mount Meru rising up at the center of the universe.

Buddhism reached China during the Three Kingdoms period (220–280 C.E.) and with it came the stupa. The Chinese adapted and interpreted the stupa to suit their culture, and built the towers with succeeding upwardly curved tiled roofs that have come to symbolize Chinese architecture. The earliest pagodas were square, and some may have been built of wood, but the oldest surviving pagodas are built of burned brick. By the eleventh century they had evolved into octagonal brick structures, representing the eight compass points of Tantric Buddhism.

There are both hollow and solid pagodas. Solid pagodas were stronger but could be climbed only externally. Eventually, pagodas acquired balconies and verandas.

Chinese brick production, like Chinese pottery, met high and exacting standards. Bricks were molded in both box molds and open molds. Fancy bricks were stamped with designs. Pagoda roof brackets were specially made of burned brick molded like traditional wood brackets. By 1103 C.E., during the Song dynasty, strict building codes were written and published, including directions for firing the bricks in reduction and specific sizes for specific purposes.[20]

The Romans, enthusiastic architects and civil engineers, embraced not only fired bricks, but also stone, especially travertine marble, and concrete for their building projects. They liked to combine materials and to exceed known possibilities. They faced brick structures, such as the Colosseum, with marble, and filled double brick walls of others with concrete. They mastered the art of building brick arches and vaults, which they used in aqueducts, amphitheaters, temples, markets, apartment buildings, and estates and sometimes just because they liked the way they looked.

The Romans were the first to make specially molded wedge-shaped bricks for building arches. Arches can be built with straight bricks, by pressing the bottoms of the bricks together to form the curve of an arch, while leaving the tops apart. The mortar between the bricks is then wedge shaped. Earlier builders had learned to cut stones into wedge shapes for making arches. And mud bricks had been used for domes and arches, especially the corbelled (or stepped) arch. But shaping the bricks into standard wedge shapes prior to firing was a great advance and made arch building easier and more reliable.

When I built my latest kiln, I used wedge-shaped bricks to form the double-layered sprung arches of the kiln roof. I am grateful to the Romans for their invention. The roof is strong and was not much trouble to make.

Roman soldiers made and used fired bricks in the far-flung lands of the empire, introducing them to places such as the British Isles where only wood, stone, and mud had been used prior to the soldiers' arrival. Wealthy owners of country estates brought in extra income by setting up brick-making operations on their farms. But when the Roman empire fell, brick making in the northern provinces all but disappeared. It did not reappear until the twelfth century.

Tiles have been used in buildings for thousands of years, most effectively in the Near and Middle East. Though made of fired clay, tiles do not play a load-bearing role in architecture. They are used for roofs, as we have seen, for floors, to clad interior and exterior walls, on kitchen counters, on the faces of stoves, benches, and tables, on stair risers, as surrounds for door jambs and fireplaces, for signs, as decorative accents, and in gardens. They can be flat or embossed, square, rectangular, hexagonal, octagonal, star shaped or free-form, glazed or unglazed. Tiles can be plain solid colors or kissed only by the smoke and flames of the kiln fire. They can be set in geometric designs as simple as a checkerboard or with a mathematically computed pattern of repetitions. They can be joined together to form a panel or frieze so that each tile contributes part of a whole image, one with the stem and leaves of a flower, another with the flower head, and two others with the halves of a basket. Tiles are long lasting, durable, easy to clean, hygienic, and can be made to suit every taste.

Tiles are adornment.

Like bricks, early tiles were cut from slabs of clay or formed in wooden molds. They were often made of drier clay than bricks as they were thinner, and might be pounded into the mold. Hans E. Wulff, in *The Traditional Crafts of Persia*, described an Iranian potter making tiles in the same way they had been made for generations:

"The tilemaker usually works in the open air. His assistant pre-pares suitably sized clay lumps. The master has a wooden mould in front of him and throws the clay lump into it with verve, beats it with his bare hands to force it into the remote corners of the mould, folds the surplus up, and cuts it away with a wire. Then he empties the mould with a swift movement. The assistant takes the tiles into the shade for the first state of drying and when they have sufficient strength he places them in a well-ventilated drying room, facedown on the flat floor, for the slow and final drying."[21]

Egyptians used tiles as early as c. 2780–c. 2680 B.C.E. Tiles and glazed bricks were used in Ashur, Nimrud, and Babylon.[22] But it was not until the rise of Islam that tile became an important architectural feature.

By the middle of the eighth century of the Common Era, Islam had spread out from Arabia and reached from Spain across northern Africa, through Egypt, Arabia, and Persia to the borders of India and China. As in the past, houses were made of sun-dried mud, but important buildings such as mosques and palaces were more and more often made of burned bricks, especially in regions where stone was scarce such as present-day Iran and Iraq. Here, bricks, usually square, were laid with fast-setting gypsum mortar in intricate pat-terns. Many of the mosques were built with vaults and domes. Instead of pagodas and stupas, Middle Eastern Muslims built minarets near their mosques, soaring brick towers, richly orna-mented, from which the faithful were called to prayer.

By the ninth century, glazed tiles were being used on the exteri-ors of these burned brick mosques and palaces. They continued to be used more and more extensively until by the beginning of the fourteenth century, the brick walls of buildings were completely hid-den beneath brilliantly glazed tiles. When the Spanish ambassador visited Timur (1370–1405 C.E.), the fierce but artistically ambitious ruler of central Asia, he was dazzled by the ruler's palace. "We saw indeed here so many apartments and separate chambers," he said, "all of which were adorned in tilework of blue and gold with many other colours, that it would take too long to describe them here, and all were so marvelously wrought that even the craftsmen of Paris, who are so noted for their skill, would hold that which is done here to be of very fine workmanship."[23]

The same trends that influenced Islamic pottery influenced Islamic tiles. Early tiles were lead glazed in bright, runny colors. In the ninth century, potters turned out shimmering luster glazed tiles with swirls of flowers, vines, prancing deer, leaves, and birds. Later tin and antimony were used to make smooth white glazes suitable for decorating with swirling lines of vivid cobalt blue. Polychrome underglazes and overglazes in dazzlingly rich hues appeared. Solid-color tiles covered the roofs of domes, shining bluer than the blue of the sky. As the decoration on tiles gained in importance, the tasks of painter and potter were often separated.

By the twelfth century, the use of tiles had exploded throughout the Islamic world, brightly covering walls in Anatolia, Turkey, Syria, Iraq, Iran, and North Africa. Tiles were everywhere.

So too were tile mosaics. The Romans made mosaics of little cubes of stone. Now mosaics were made of clay tiles, cut or broken after they emerged from the kiln. These small irregular pieces of glazed and fired ceramic were laid in mortar on floors, panels, and walls, giving a rich, detailed look that could not be achieved by larger tiles. Tile mosaics covered the domes of mosques and minarets.

As northern Europe came out of its sleep, builders began to use ceramic tiles, usually unglazed, fire kissed, on the floors of stone and brick churches and castles. The Cistercian monks made encaustic floor tiles, tiles with designs created by using different colored clays all the way through the tile so that the image would never wear away.

Soon, influenced by the glazed tiles coming from the southern regions, northerly European potters were also turning out pretty wall tiles and tiles for stoves. Potters working on the outskirts of cities in France, Spain, and Italy found the demand for tiles increasing and kept busy supplying wealthy clients.

Fired bricks began to be produced in vast quantities throughout Europe. In Italy, brick makers kept permanent kilns that they fired from spring until the autumn. In England, brick makers were itinerant and made bricks on location with clay dug and fired on-site.

An early-nineteenth-century English family work together to make bricks.

By the thirteenth century, the bricklayer's art had reached new heights in Europe and imposing burned brick buildings arose. In Florence in the early fifteenth century Brunelleschi built his famous dome over the cathedral of Santa Maria del Fiore using millions of bricks. Here, confronted with the task of covering a cavernous span, and unable to use the fast-setting gypsum mortar that the Persians used for their domes because it would not withstand the Italian rains, Brunelleschi solved the problem by standing bricks on edge at intervals, so that they were vertical, thus keeping the dome from collapsing in upon itself.

When the Industrial Revolution swept across the continent, tile and brick making like other ceramic activities entered the factory. New processes were invented, including dust pressing, in which tiles were made dry, and the ram press, which used force to produce tiles. Prices dropped and the use of tile and brick became more widespread.

Pug mills to mix clay were one of the earliest attempts at mechanization during the nineteenth century, though they often employed the muscle of a draft horse. Various brick-making machines were also patented. In 1793 a Connecticut man, Apollos Kingsley, patented a machine with a charger to compact clay into bricks. In 1845 a man named Alfred Hall patented a combination pug mill and brick press which required both the energy of a horse and a man to operate. Though colonists had made and fired bricks for their houses from clay dug in their own fields, especially in the southern states, Americans, always thrilled with a new invention, were far more interested in automating brick- and tile-making processes than Europeans. Eventually, extruders to squeeze out hundreds of stiff mud bricks came into common use.

Two horses provide the "horsepower" for a mid-nineteenth-century pugmill used to prepare clay for local potters or brickmakers. Similar pugmills were used on both sides of the Atlantic.

Throughout history, bricks have been fired in "clamps" as well as kilns. A clamp, or scove kiln, is essentially a kiln built of the unfired bricks themselves. As the fuel burns, the bricks are baked. Bricks were also fired in updraft kilns and downdraft kilns. In the late nineteenth century, after the invention of the continuous kiln, brick makers turned more and more to the twenty-four-hour kiln. Today, though burned bricks can still be made and fired by hand, brick making in much of the world is an industry operated by huge, often multinational, corporations. Brick sizes are standardized with slight differences from country to country. They come in the natural colors of burned clay, red, yellow, and tan as well as dyed or glazed. Some are made with hollows to aid the construction process. Usually, they are stamped with the factory name.

Nevertheless, bricks themselves remain relatively unchanged since ancient times with the exception of firebrick. Firebrick is specially formulated with refractory clays to withstand high temperatures. Insulating, or "soft," firebricks are lightweight, porous, and do not transmit heat the way regular, dense, hard firebricks do. Hard and soft firebricks are used in everything from household heating systems to artist's kilns to industrial furnaces. They increase efficiency, withstand high temperatures, and, in the case of hard firebricks, are resistant to corrosion.

Ironically, though one of the wonderful qualities of clay is that it can take virtually any form, the principal building material made of clay, the brick, is a rectangular box shape with sharp edges and crisp corners. Nevertheless, there have always been builders and architects drawn to the malleable qualities of mud.

Antonio Gaudí i Cornet (1852–1926 C.E.), a Catalan architect, rejected the cubes and right angles of traditional building, yet used brick, tile, and mosaic to create his organic, curvilinear structures. Gaudí's roof tiles look like the scales of a fish's back. His window openings and bell towers appear elastic. His turrets make us think of jewel-encrusted onions. Gaudí's buildings surprise us, even a century after they were made.

Two architects, Ray Meeker in India, and the Iranian Nader Khalili, who now lives in California, have experimented with firing houses whole, after they are built. Khalili has successfully fired existing mud-brick houses. He has also built houses of adobe bricks without organic matter so that once fired, they are as strong as prefired bricks. Meeker has built large potlike houses in India and fired them. Both men believe that houses have a spiritual aspect to them and have tried to find a way to make affordable long-lasting ceramic housing. Though fired, their buildings have a soft, sculpted quality.

If you walk around New York City you will also begin to notice, here and there, sculpted clay amid the glass and steel and concrete of skyscrapers. Look up at the buildings and you will discover brightly glazed bands of tiles, lions, buffaloes, flowers, medallions, cornices, columns, griffins, Aztec gods and symbols, and leafy swags, all made of heavily grogged terra cotta. In the midst of the pollution and grime of the city appear glazed ornaments of clear blues, greens, purples, yellows, and bright reds.

The Flatiron Building, one of the city's best known structures, is a showpiece of architectural terra cotta. "Above a stone base, the building's curtain wall is a combination of brick and ornate terra cotta. Striking ornament appears on the fourth floor, where ovals with foliage alternate with rondels of a somber woman's face. The elaborate overhanging cornice and upper stories have wonderful details including lions' heads and mask-like faces,"[24] writes Susan Tunick, president of the Friends of Terra Cotta.

Terracotta began appearing on buildings in New York during the latter half of the nineteenth century. Clay ornaments had long been successfully used in England and the European continent, but it was new to America. Before it came to New York, clay was used in Chicago after the Great Fire. To their consternation, Chicagoans learned that unprotected iron is not fireproof, and when rebuilding got under way, they began guarding their iron beams with clay. They also made fireclay tile ceilings as a protection against another conflagration and even substituted clay work for stone. Clay lintels and cornices were a natural progression from these structural uses, and soon ornaments were being made.

New York architects were at first skeptical that terra cotta could withstand the rigors of the city's winters, but once they were proved wrong, terra cotta was used on many buildings. And brick continued to play a role. Most of the early skyscrapers were built of brick, or brick covered with stone. Three million white glazed bricks were made for the Chrysler Building. The Empire State Building includes ten million bricks covered with stone.

The role of clay as a building material has continued. Tiles were key design elements during the Art Nouveau, Arts and Crafts, and Art Deco movements. Pink and aqua bathroom tiles were a fad of the 1950s. Today, big hexagonal and square Mexican earthenware tiles cover the floors of "gourmet" kitchens in upscale houses. Towns build their government buildings and schools of brick. American colleges are predominantly brick. Mosaics decorate the public restrooms of subways and train stations. And in much of the world, adobe is the building material of choice and necessity.

We may be the only species that makes dishes and ovens of clay, but we are not the only species to build our houses of clay. Mud dauber and potter wasps use clay to build their homes. The daubers' nests, shaped like organ pipes, are of such pure clay they can be fired in a kiln. The potter wasps' nests look like bottles. And some birds make nests of wattle and daub, sticks and mud. With the recent discovery that water was once on Mars, we have to wonder if there was clay too, and if so, did the "Martians" use it to build their houses?

SANITATION
A Nice Hot Bath, a Drink of Water,
and Don't Forget to Flush

You will be scarp'd out of the painted cloth for this: you lion that holds his Pollax sitting on a close stoole, will be given to Ajax.

—WILLIAM SHAKESPEARE,
Love's Labour Lost

QUEEN ELIZABETH WAS embarrassed; everyone in England was talking about the racy book her godson Sir John Harington (1561–1612) had illegally published, *A New Discourse on a Stale Subject, Called the Metamorphosis of Ajax*. For the second time in his short life, she angrily banished him from his home and the court.

Harington was part of the extended royal family, though not actually related by blood. The queen made him her godson because his father had been married to the illegitimate daughter of Henry VIII before he married his second wife, Harington's mother. Harington, charming and accomplished, alternately endeared himself to his godmother and horrified her. While still in his twenties, he translated the racy parts of *Orlando Furioso*, the epic poem by Ariosto. The straitlaced queen was furious. She decided to teach the young man a lesson and sent him away with orders not to return until he had translated the entire poem—all forty-thousand lines of it! Harington, a talented poet in his own right, was delighted with his punishment and immediately set to work on the translation.

When his task was completed (it is the best-known English translation even today) the queen let him return to the court, and for five years he behaved, at least in matters literary. But Harington was irrepressible and soon he wrote the "dirty" book that upset the queen, *The Metamorphosis of Ajax*. The book was filled with double entendres and obscenities; "ajax" was a play on "jakes," slang for chamber pot. Harington's main point, however, was to criticize and poke fun at his fellow countrymen for their unclean habits. He wrote that he hoped that his readers would not find his treatise "noysome and unsavory,"[1] and then presciently argued that filth and infection were linked, and that contemporary English living conditions were unhealthy.

He was correct.

Daily life during the Middle Ages was rife with foul odors, disease, and filth. Peasants lived in thatch-roofed hovels with dirt floors. Garbage was left in a corner or tossed out the front door. Tenement housing in the cities was crowded and dark. Landlords were taxed on the windows of a building and so were reluctant to include more than a few. Baths were rare. At the heart of Harington's book was his invention and solution: Captain Ajax, the flush toilet. He included diagrams, instructions, materials lists, and costs. "This device of mine requires not a sea full of water, but a cistern, not a whole Thames full, but half a ton full, to keep all sweet and savourie,"[2] he wrote. Modern toilets are surprisingly similar to Harington's in concept. His had a valve at the bottom of the water closet and a system for washing down the waste.

Though Harington was denied a license to publish his book, it was illicitly printed three times and became a bestseller. However, it was the salaciousness of the work that attracted readers rather than the promise of the flush toilet—no one believed Harington's premise that a fouled environment caused disease.

Eventually, the queen relented and let Harington return home. He convinced her to let him build a flush toilet for her personal use

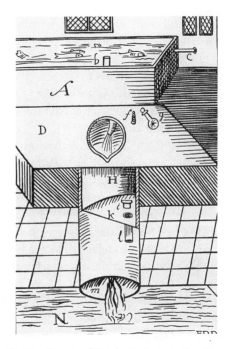

Working diagram of Sir John Harington's Don Ajax.

in her Richmond palace. Though she was not terribly comfortable with the device, she kept a copy of her godson's book chained to a nearby wall and did use the flush toilet now and then. But Harington was unable to persuade anyone else to build his device. Only he and the queen had the appliances installed in their homes.

Consequently, until a few enterprising potters reinvented and perfected the flush toilet two and a half centuries later, thousands of people needlessly perished, and by the early nineteenth century, a deadly cholera epidemic raged from Asia through the Middle East, Europe, and North America.

As long as human populations remained sparse, it was safe for sanitation practices to be casual. If you were hot and sweaty, you could take a dip in a river or pond. If you needed to relieve yourself, you could walk away from your tent or hut and do your business behind a bush. You could even move your encampment to fresh territory.

Throughout much if not all of prehistory, people likely preferred to defecate away from their living areas. Later, they wrote down rules about where to relieve oneself. Deuteronomy (23:12) tells the Hebrews,

> You shall have a place outside the camp and you shall go out to it; and you shall have a stick with your weapons; and when you sit down outside, you shall dig a hole with it, and turn back and cover up your excrement.

Aristotle gave a similar prescription, and told the boy Alexander the Great to dispose of animal and human waste a distance from camp.

With the rise of cities came the need for drains to remove waste, and pipes and wells for fresh water. Though the ancients did not understand that there were invisible organisms in contaminated water that could lead to sickness and death, they did understand convenience, and had some notions of the connections between dirty water and ill health. Hippocrates (c. 460–c. 360 B.C.E.), the "father of medicine," worried about visible impurities in water and suggested that it should be boiled before drinking.

One of the first sanitary conveniences was the chamber pot. The Romans credited the Greek Sybarites, the notoriously luxury-loving inhabitants of the city of Sybaris on the Gulf of Taranto, with inventing the chamber pot. Perhaps the idea came first to a reveler who cleverly—or mistakenly—urinated into an empty cooking pot. Perhaps it was an innovative potter who first thought of making specialized pots for the purpose. However it originated, the idea caught on: potters made and decorated terra-cotta pots to be used specifically for relief. The hedonistic Sybarites began taking these terra-cotta chamber pots to their parties.

The ancient Romans also used chamber pots, usually a plain clay jar, which they emptied out a window or into a public cesspool. Like the Greeks, they brought their chamber pots to parties and, drinking and eating to excess, vomited and urinated into them. Petronius, a first-century Roman courtier, explained in his writings that "when a gentleman wanted his chamber pot" he would "make a noise with the finger and thumb by snapping them together."[3]

At parties, drunken guests inevitably lost their balance, stumbled, and broke their chamber pots or the pots of other guests. The shards of these pots, strewn about the floor the morning after, were seen as an indication of a successful banquet. Sometimes a guest would be so drunk he used the staircase or other out-of-the-way spot to relieve himself. Revelers too poor to bring a chamber pot also used the stairwells. This led to the expression "He doesn't have a pot to piss in."

If you were sober enough to walk home, you could always make use of the vessels left at street corners for public relief. There was no charge for using one of these early public rest areas, but the emperor Vespasian (9–79 C.E.) decided to enrich his treasury by charging rent to the fullers who emptied the jars into larger vats and used the urine in the dyeing process.

Although they liked to bring their crockery chamber pots to parties and dinners, the Greeks and Romans, and many ancients before them, also enjoyed more advanced sanitary amenities, amenities that, by the Elizabethan period during which Harington lived, had been lost in the Dark Ages, or earlier forbidden by Christians seeking to distance themselves from the "sinful" ways of the Romans. The availability of clay, its workability and durability, together with the increasing skills of potters, enabled the ancients to develop elaborate systems of sanitation and, for the wealthy, a comfortable lifestyle.

The city of Olynthus, in northern Greece, boasted ceramic tiled bathrooms with bathtubs. The tubs had drains and emptied into underground clay pipes. Greek women kept fancy portable terra-cotta tubs in their homes. "Health nuts," the Greeks were very interested in strong bodies and strong minds and worked hard to maintain a robust physical and mental fitness. By the seventh century B.C.E., most Greek cities offered public gymnasiums with hot and cold water. Men, however, eschewed hot water baths as indulgences of the faint-hearted. They preferred the rigors of ice cold water. Women felt no such compunction and liked to stretch out in soothing baths of heated water.

Hippocrates urged his patients and fellow citizens to sit upright in their tubs for good health rather than recline or lie down. And he too extolled the virtues of freezing cold baths. A cold bath was often taken standing up, actually more a shower. Perhaps sitting in a tub of cold water was too much even for the athletic, health-conscious men to endure. A slave would douse the bather with frigid water from a pottery jar. Good hosts, affluent enough to own a house and slaves, offered such a refreshing dousing to their travel-weary guests.

One hundred thousand people lived in Knossos, the capital of the Minoan sea kings (3000–1100 B.C.E.), on the island of modern-day Crete. Because the island was racked with earthquakes, the capital had to be reconstructed several times. Eventually it grew into an impressive city, covering twenty-two acres. It was here that some of the earliest and most remarkable advances in sanitation were made.

During the reign of King Minos (c. 1700 B.C.E.) the palace was extensively rebuilt. By 1500 B.C.E. it sprawled over five acres and consisted of fifteen hundred rooms and corridors. Earthenware sewer pipes, made by local potters, were laid beneath the corridors. The pipes were formed in 2½-foot sections, with one end tapered down to an inch in diameter and the other end widened. The pipes were fitted together with the narrow end of one pipe inserted into the wider end of the adjoining pipe, creating a relatively tight seal. Four cement-lined limestone shafts served these sewers from the rooms above, for both ventilation and the disposal of waste.

The palace was also outfitted with rooftop cisterns for the collection of rainwater. The water flowed through terra-cotta pipes for use in the rooms below. Some of the water flowed constantly into the king's flushing toilet and washed the waste into the sewer pipes. The queen enjoyed a lavishly decorated earthenware tub in her private bathroom. It was a sumptuous five feet long but had to be filled and emptied by hand because there was no outlet.

Clay water pipes, for drainage and for irrigation, were also used in Babylon and Egypt. The clay was tempered with straw and fired in kilns. The Egyptians and Babylonians drew water from their wells with porous earthenware jars. The jars were attached to great wheels. As one jar was lowered into a well by slaves, the other was pulled up. Similar waterwheels can still be seen in the Middle East today, though animals power them.

It was the Romans who brought plumbing to great heights and who made bathing a social activity. Extravagant public buildings with mirrors, marble steps, statuary, vaulted ceilings, and mosaics, the baths offered a choice of bathing rooms with different temperatures, from the frigidarium (cold) through the tepidarium (moderate or tepid) to the thermae (hot) and the caldarium (very hot). Furnaces were fired in rooms adjacent to the baths. The heat was transported beneath the bath floor through air channels created by upright clay bricks. It was also brought into the room via hollow clay wall tiles.

At first, men and women bathed separately, but in time they were allowed to bathe together as long as they didn't "stare." Finally, they not only "stared," but also ate and drank and played together in the baths. The small rooms, the bordellos, began to take on another meaning, at least in the minds of the disapproving early Christians. Warning believers against bathing, St. Benedict wrote, "All is vanity to those that are well, and especially for the young, bathing shall seldom be permitted."

The Romans spread their ideas throughout their vast empire, which, at its height, included all the lands encircling the Mediterranean Sea, Mesopotamia, most of modern Europe, and into Great Britain. They built aqueducts, sewers, and baths. Aqueducts provided a continuous flow of water into homes. Underground sewer pipes, originally made of clay as in earlier cultures, were later made of lead, molded in clay. Cisterns provided water to flush the indoor toilets of the wealthy, conveniently positioned near the kitchen oven for the women to use not only for their relief but for cooking refuse. Men had their own toilets, often outside the house.

But as the Roman empire crumbled, as the Romans retreated, their ideas of hygiene, and their habits of bathing, fell into disrepute. The great baths at Bath, which the emperor Claudius built on twenty-three acres in Britain on the outskirts of the empire, were abandoned after five hundred years of Roman occupation. The same happened throughout the former empire.

Meanwhile, the humble chamber pot remained a personal necessity. Since ancient times, it had been customary to fling the contents of a chamber pot out a window or into a common cesspit. On occasion, the whole pot would be flung out into the street, endangering any pedestrians passing by. Aeschylus, the Greek dramatist, called these flying chamber pots "missiles of mirth," though one wonders just how mirthful it was to be struck by the pot or its contents. The Romans attempted to legislate this behavior and fined anyone who injured a passerby with a chamber pot or its contents flung from a window during the daytime. Apparently the law did not protect evening strollers.

Laws or not, manners dictated that the person flinging the contents of a chamber pot shout out a warning. The French yelled "gardez l'eau," which the English changed to "gardyloo." The Italians hollered out euphemistically, in Italian, "Take away your lantern." In 1395, Parisians outlawed throwing the whole chamber pot out the window, and by the seventeenth century, they banned throwing the contents out. The authorities could not think of everything: in 1418 a Parisian man lost his temper and smashed a chamber pot over the head of a lady. He was quickly exiled.

An unsuspecting medieval stroller is drenched with the contents of a chamber pot.

Medieval chamber pots were robustly thrown earthenware with a lead-glazed interior and often a splash or drip of glaze on the exterior (though later some chamber pots were made of tin). They usually had one handle and were left undecorated. Today, archaeologists sometimes have difficulty distinguishing between a pot used for cooking porridge and a chamber pot. Called "originals" or "jerseys," these rustic household pots were kept out in the open in plain sight of family and guests.

By the sixteenth century, the wealthy began to have their chamber pots made of silver or gold or even glass. During the eighteenth century, porcelain replaced earthenware as the most common material. The pure white porcelain was covered with clear glaze and exuberant decorations, including flowers, garlands, scenery, and frequently humor. Emblazoned on the Duke of Wellington's chamber pot was the verse,

> *Keep me clean and use me well,*
> *And what I see I will not tell.*

The French apparently did not regard the idea of what a chamber pot might "see" as funny, whether or not the pot "told." In the mid nineteenth century, a French potter was arrested for embellishing the interiors of his chamber pots with a large eye and the words, "I see you."

However, the French did not completely eschew humor in the decoration of their chamber pots. When Louis XVI discovered that one of his lovers had a crush on the visiting Benjamin Franklin, he ordered a chamber pot with a cartoon of the American and sent it to her as a little surprise gift. Pots with Franklin's likeness suddenly became all the rage in Paris.

Potters used the transfer printing technique to depict daily life (a mean-looking woman wielding a poker on her cowering husband, a sailor gone off to sea, fiddlers, sailing ships, and numerous verses, sayings, and jokes). In Northern Ireland, images of Prime Minister Gladstone appeared inside chamber pots, a protest against his home rule policies.

Ironically, as chamber pots became more decorative, they were also more likely to be kept hidden. They were set in chairs (the original potty chairs) and stashed in cupboards (commodes). Some were kept in closets. Many of the great furniture makers, Chippendale, Hepplewhite, and Sheraton, produced pieces to hide or camouflage the humble chamber pot.

But no matter how ornate the pots, or where they were kept, the problem remained: what to do with the contents. Sometimes, they simply spilled. The Dominican monk Felix Faber of Ulm, wrote in *A History of a Private Life*, a description of his trip to the Holy Land from 1480 to 1483, "Each pilgrim has near his bed a urinal—a vessel of terra cotta, a small bottle—into which he urinates and vomits. But since the quarters are cramped for the number of people, and dark besides, and since there is much coming and going, it is seldom that these vessels are not overturned before dawn."[4]

More often the contents were dumped into open sewers, the street, public cesspits, and, in London, the Thames and Fleet rivers. It was not uncommon for a neighbor's cesspit to overflow, or seep into the house next door. In 1189, Londoners enacted rules to govern how far a cesspit must be from a property line, but the rules were often ignored. Stench and seepage were rampant. Those well off enough to live near the riverbanks built their houses with a garderobe, a small latrine that cantilevered out from the house, with an opening for the waste to drop directly into the water below. In castles, the waste dropped into a moat, onto rocks, or sometimes into a pit in the cellar.

One of the highest paid jobs was that of "raker" or "gongfermor." These men cleaned out the moats and cesspits and cleared the ditches. In 1326 Richard the Raker perished when the rotting wood of his privy gave way and he plunged into the depths of his own cesspit.

Well into the nineteenth century, living conditions for much of the populace were foul. The reeking sewers and cesspits were a source of disease and complaint.

Appalled, the journalist Henry Mayhew wrote in 1874 in his *London Sketches*:

As we gazed in horror at this pool, we saw drains and sewers emptying their filthy contents into it, we heard bucket after bucket of filth splash into it, and the limbs of vagrant boys bathing in it seemed, by pure force of contrast, white as Parian marble. And yet, as we stood gazing in horror at the fluvial sewer, we saw a child lower a tin can, with a rope to fill a large bucket which stood beside her. In each of the rude and rotten balconies, indeed, that hung over the stream, the self-same bucket was to be seen in which the inhabitants were wont to put their mucky liquid to stand, so that they might, after it had been left to settle for a day or two, skim the fluid from the solid particles of filth and pollution, which constituted the sediment. In this wretched place we were taken to a house where an infant lay dead of cholera. We asked if they really did drink the water? The answer was they were obliged to drink the ditch, unless they could beg or thieve a pailful of the real Thames.[5]

Earlier, the novelist Tobias Smollett (1721–71) described the prevalent filthy conditions in Scotland in *Humphrey Clinker*:

And now dear Mary, we have got to Haddingborrough, among the Scots, who are civil enuff for our money ... But they should not go for to impose upon foreigners; for the bills in their houses say that they have different EASEMENTS to ley; and behold there is nurro geaks in the whole kingdom nore anything for pore servants, but a barrel with a pair of tongs thrown across; and all the chairs in the family are emptied into this here barrel once a-day; and at ten o'clock at night the whole cargo is flung out of the back windore that looks into some street or lane, and the maids call GARDY LOO to the passengers, which signifies LORD HAVE MERCY UPON YOU! And this is done every night in every house in Haddingborrough; as much as you may guess, Mary Jones, what a sweet savour comes from such a number of perfuming pans.

■

WITH SUCH FILTH and contamination, widespread and catastrophic disease was inevitable. The first wave of the Black Plague, though not caused by leaking cesspits, open sewers, or polluted rivers, was enhanced by the garbage and muck in the streets. It struck England in 1348, killing thousands.

Other horrific diseases, which were linked to the contamination caused by the open sewers, the contents of chamber pots, and the human excrement in the rivers, such as dysentery, typhus, typhoid, and cholera, struck the populace. The death tolls were tragically heavy.

Dysentery killed many Crusaders, perhaps more than died by the sword. Typhus spread by lice, mites, and fleas in human feces. It ravaged men who lived in close quarters (it was often called ship fever or jail fever), exposed to one another's chamber pots, unwashed hands, and feces-contaminated drinking water. Napoleon lost many of his soldiers to this painful and deadly disease. Typhoid fever, recognized as a distinct malady in 1837, killed forty-three thousand British soldiers in South Africa after they drank water from a river contaminated by human urine and feces. Typhoid epidemics swept through the Americas, killing many.

But worse than typhus or typhoid was cholera, the first truly global disease. Cholera is spread by contact with the feces of a cholera victim. The feces can be in the drinking water, in a chamber pot, on clothing or bedding, on unwashed hands.

Cholera festered for years along the Ganges River in India. Then, in 1817, after a three-month festival that attracted pilgrims from great distances, it broke out of India. Untold numbers of Indians living near the river died. Ten thousand British soldiers died along with them. Pilgrims and merchants returning home from the festival spread the disease along the trade routes to Iran, Russia, and eventually Europe. Cholera thrived on crowded ships and soon reached the United States and Canada where by 1832, it was epidemic in New York City, Montreal, Quebec, and Detroit.

North America was less populated than England or Europe, and though chamber pots were used, and the contents were often flung out the window, European style, the sanitary problems were not quite as grave. Americans built outhouses, both rough-hewn wooden structures and formal brick structures, in their backyards and in public places such as parks and schools. But these offered little protection against the dangers of polluted water. Outhouses were often sited uphill from the local wells and contaminated the drinking and washing water in even rural areas. They were no defense against the spread of cholera.

Officials argued about the best course of action. Quarantines were imposed upon English seaports, but merchants and bankers railed against the embargoes and the financial devastation they feared would result, especially in the lucrative international textile business. Many physicians believed the quarantines would be ineffective and sided with the merchants.

The empty streets of New York were sprinkled with chloride of lime while the sick and dying stayed inside. The militia in the Northeast attempted to prevent immigrants from entering the United States by confining them to their ships, but the resourceful immigrants, desperate to leave the dank and disease-ridden vessels, leapt overboard and swam to shore, spreading cholera throughout Vermont and upper New York State.

Death from cholera was often swift. Victims vomited, had high fevers, and suffered devastating diarrhea. Their whole bodies were racked with muscle pain. The end could come as quickly as within twelve hours of exposure, though some victims lingered for as long as forty-eight hours.

The German poet Heinrich Heine described the situation thus:

A masked ball in progress … suddenly the gayest of the harlequins collapsed, cold in the limbs, and underneath his mask, violet blue in the face. Laughter died out, dancing ceased and in a short while carriage-loads of people hurried from the Hotel Dieu to die, and to prevent a panic among the patients were thrust into rude graves … Soon the public halls were filled with dead bodies, sewed in sacks for want of coffins.[6]

By the end of the first epidemic, medical personnel and government officials began to realize that there was a relationship between poor drainage, human crowding, and cholera. In 1854, a British doctor, John Snow, traced the source of five hundred cases of cholera to sewage-polluted water. Two decades later, Dr. Robert Koch was able to isolate and identify the bacillus of cholera.

The heavy stench of grossly polluted rivers on hot summer days, open sewers clogged with offal and fecal matter, leaky cesspits, filthy streets, dirty chamber pots left to stand for hours or emptied out windows made life increasingly difficult and unpleasant. Even before the connection between contaminated water and disease, between hygiene and health, were understood, there were cries for reform.

For thousands of years, what sanitation was practiced was largely reliant on ceramics. But now, with increasingly dense populations, world travel, and sweeping epidemics, these ceramic products—washbowls, chamber pots, earthen pipes—were no longer adequate for the task. At the height of these epidemics, industrial potters made two lifesaving contributions to sanitation. They offered enormous progress in the elimination of diseases such as typhus, typhoid, dysentery, and cholera and vastly improved the quality of life.

In the mid nineteenth century, Henry Doulton (1820–97), with his friend Edwin Chadwick, began arguing against the use of open sewers and earthenware sewer pipes in London and in favor of vitreous stoneware sewer pipes. Effluence would travel more efficiently through nonporous stoneware, Doulton explained. The pipes would be cleaner. And an enclosed underground system would be far more sanitary than the open sewers and rivers of the day. And he would be happy to produce the necessary stoneware pipes in his factory.

Henry's father, John, was part owner of the Doulton & Watts pottery, where he made stoneware bottles for beer, chemicals, and other liquids. Henry apprenticed in the pottery at the age of fifteen, and quickly demonstrated his skill and ingenuity. Though very young, he expanded the factory's product line to include garden ornaments and building materials. From these successes, he prevailed upon his father to enter into the stoneware sewer pipe business. Fortunately, the case he and Chadwick made to the London city officials was convincing, and the Doulton factory was awarded a contract to produce vast amounts of pipe. The clay sewer pipe the

Doultons made and installed in London in 1848 greatly improved the conditions of the city and is still in use today!

Doulton was not alone in his efforts. Vitreous clay sewer pipes were installed beneath other cities. According to the National Clay Pipe Institute, pipe was installed in Washington, D.C., in 1815; in Boston, Massachusetts, in 1829; in Sydney, New South Wales, in 1832; in Manchester, England, in 1845; and in Edinburgh, Scotland, in 1850. These pipes are also still functioning today.

The second great contribution of potters that improved sanitation and controlled disease was the reinvention of the flush toilet, this time made of clay. A clockmaker, Alexander Cummings, started things when he invented a slider valve for his new toilet. When it worked, it prevented the noxious cesspit gases from entering the room. Unfortunately, the metal valve rusted and often stuck, creating the very problem it was designed to solve. Cummings also used a valve to let water into the bowl to cleanse it. This too sometimes stuck so he provided for overflow (a rather unpleasant thought). Cummings received a patent for his inventions in 1775.

Three years later, locksmith and engineer Joseph Bramah partnered with the great industrial potter Josiah Wedgwood to make a better flush toilet. Bramah replaced Cummings's slider valve with a hinged outlet valve and obtained a patent for his design. Wedgwood made bowls for Bramah and for several other manufacturers too, though he was paid little for his work. These early water closets were discretely hidden in fancy mahogany cabinets, much as the chamber pots of the wealthy.

Then in 1782, potter John Gaitlait decided that he could do more than merely produce clay bowls for toilets designed by valve manufacturers. He would take charge of the whole process himself. Gaitlait designed and made an all-clay flush toilet with a unique water-sealed U-bend stink trap. There were no metal valves to rust or malfunction—a remarkable design breakthrough.

Soon other established and well-known potters, including Doulton, Copeland, Minton, Shanks, Wedgwood, Twyford, Edward Johns, and Enoch Wood were also working directly with plumbers to design and manufacture sanitary toilets, washbowls (sinks), and bidets.

Still, the toilets were made in two pieces, the bowl and the pedestal, and were enclosed in traditional wooden cabinets. Gases and odors could leak from the joint of the two pieces.

Finally, potters began to eliminate the wooden cabinets so that problems could be more quickly detected and repairs more easily made. The early ceramic toilets were fashioned of the same white earthenware clay as tableware. In order to garner public acceptance for these all-clay, cabinetless toilets, potters decorated the bowls and pedestals with embossed designs and colorful images: flowers, vines, and dolphins.

One of the leading manufacturers of flush toilets was the pottery run by the Twyford brothers, Thomas and Christopher. Sensing market opportunity, Thomas had moved the family firm away from domestic pottery and into the production of sanitary ware in 1848. Three years later, he collaborated with the entrepreneur George Jennings and installed one of his early toilets in the Crystal Palace and charged a penny to use it. The public was wildly enthusiastic about Jennings and Twyford's pay toilet, and Jennings placed many more around London. They too were popular and Jennings collected a considerable amount of money from them. In fact, he never charged the city for the installations, as the fees he collected accumulated so quickly.

Thomas Twyford believed that the best improvement he could make to the flush toilet would be to develop a one-piece pedestal toilet, with the bowl and pedestal joined to form a single piece. He worked on this project with his young son Thomas William Twyford. Thomas Senior died unexpectedly at the age of forty-five, leaving Thomas the younger, at twenty-three, in charge. Young Thomas inherited not only his father's factory, but his obsession with developing a one-piece toilet. He knew the bowl and the pedestal would have to be joined in the green ware stage (unfired) and they would have to withstand the kiln fire without warping, splitting, or cracking. The goal seemed unreachable as test after test emerged from the kiln ruined. Finally, he succeeded and he produced a perfect one-piece toilet. Not content with this accomplishment, T. W. also made and placed the trap inside the pedestal during the green ware stage. His other creative idea was to make the spreader (holes under the rim of a modern toilet) an integral part of the bowl, rather than a messy copper attachment as was commonly in use. His last innovation was to make an after-flush chamber, which safely sealed the toilet against sewer gases.

Twyford named his toilet the Unitas and introduced it in 1883. Queen Victoria had the opportunity to try the Unitas in 1886 when she stayed at the Angel Hotel. She immediately had one installed in Buckingham Palace. The Queen knew in a very personal way the dangers of contaminated water, poor hygiene, and unsanitary conditions. She lost Prince Albert to typhoid in 1861, and a decade later, her beloved son nearly succumbed.

T. W. Twyford's Unitas. This rare example is a two-piece, but most were one piece.

The beautifully decorated (one design had an Oriental-influenced blue and white bowl interior, with raised acorns and oak leaves on the all-white exterior) and very functional Unitas was an immediate hit. Within the first two years of its introduction, ten thousand were sold. The grateful queen bestowed on T. W. Twyford the "Royal Warrant of Appointment as Bathroom and Washroom Manufacturer to her Majesty Queen Victoria's Government." T. W. Twyford was now the leading sanitary potter of the day.

Henry Doulton used the same strong stoneware for his toilets as he used for his sewer pipes. The stoneware was impervious to water or moisture, nearly unbreakable, and could be economically once-fired. At first he salt-glazed his sanitary ware, and banded it with blue and green raised designs, but consumers resisted the overall orange color of these toilets. He then slip-glazed his toilets with white earthenware, to give them the look of the earlier low-fired flush toilets that had come, in their whiteness, to symbolize sanitation. The public embraced these toilets.

In 1887, Henry Doulton was knighted for his work with clay sewer pipes and other sanitary ware and by 1901, his firm had adopted the name of Royal Doulton. Few collectors of Royal Doulton figurines today probably realize that the "Royal" title was bestowed for Sir Henry Doulton's contributions to the development of sewers and toilets and not for the art pottery the firm is now best known for.

Potters had enormous technical challenges to overcome in the production of their toilets such as preventing warpage and cracking during the forming and drying, and especially during the firing and getting a good glaze fit. However, once these difficulties were overcome, and the new wares could be produced reliably, demand was so brisk that the coal fires of the kilns, particularly around Staffordshire, burned constantly. The potteries produced elaborate illustrated catalogs of their sanitary ware and displayed their latest designs at shows, thus creating even more demand.

At first the water for flushing these early toilets was collected in cisterns on the roof. Later the cisterns were replaced with wood-enclosed water-filled tanks set high above the toilet, and close to the ceiling. But it was not long before potters began making the tanks of clay, and set them close to the toilet. They also replaced the iron wall lugs, which held the seat in place, with clay made as an integral part of the bowl.

The name often associated with the invention of the toilet is Thomas Crapper. Crapper (1836–1910) was a plumber and a businessman who held patents for improvements to manhole covers, drains, water closets, and pipes. He operated several Crapper Plumbing shops and, though not one of the inventors of the toilet, sold toilets with his name printed on them. American soldiers in England during World War I saw these toilets, marked "T. Crapper-Chelsea," and began calling the toilets themselves "crappers." They brought the slang home with them. Alas, the word survives to this day.

The poet Henry Wadsworth Longfellow imported a toilet from England and installed it in his home. He was the first American to do so. Toilets were also installed in several upscale hotels. Soon Americans were clamoring for toilets of their own.

In response to the growing interest, U.S. potters began making toilets, but their earthenware versions could not compare with the stoneware and porcelain of England. By 1873, the demand was so great that Twyford, Doulton, and forty-one other British potters were exporting their water closets to the United States. The exports were very expensive.

That same year, a New York state potter, Thomas Maddock, succeeded in manufacturing porcelain toilets in his workshop. Despite their weight, he packed his toilets in bags and sold them door-to-door in New York City. Other U.S. potters followed. Now with the more affordable domestic toilets available, Americans began converting their unused bedrooms into bathrooms.

Also in 1873, an Austrian immigrant to the United States, John Michael Kohler, founded his company to make hog scalders. Because of their size and shape, the scalders made wonderful bathtubs. Kohler realized the business possibilities in ceramic sanitary ware and he too began manufacturing porcelain toilets.

Changes and improvements have been made to the potter T. W. Twyford's original Unitas, including the symphonic toilet and the wash down. Modern toilets have been designed to use less water. And during the mid twentieth century, pastel colors, such as pink and aqua, were popular. But the one-piece clay toilet, invented by the industrial potters of the nineteenth century and essentially unchanged since that time, are an indispensable feature of the twenty-first-century household.

Today the manufacture of ceramic toilets, sinks, bathtubs (sinks and tubs often being ceramic-coated iron or steel), and sewer tile are the purview of big business. Continuously operated tunnel kilns, controlled by sophisticated computers, are used to fire the ware.

Heavy-duty extruders form modern sewer pipes. Nevertheless, the basic material, clay, remains the same as that used by the Sybarites for their chamber pots and the Minoans for their sewer pipes. Indeed, even the shape of today's sewer pipe resembles those of King Knossos, with the narrow end of one pipe fit into the wider, flared end of the adjoining pipe.

Reliable clay toilets, sewer pipes, sinks, and tubs prevent the spread of germs and bacteria and add to the convenience and comfort of modern life. They help protect our supplies of drinking water from contamination. It makes one smile to think that mud, that gooey, sticky, "dirty" substance, was the crucial ingredient in facilitating cleanliness for the ancients, and it is crucial for us—with our gleaming, often multiple bathrooms—today.

8

FARMING MADE EASY
Irrigation, Propagation, and Incubation

. . . loved the effects of flowerpots which decorated
the roofs and balconies of Rome.

—PLINY THE YOUNGER

I N EIGHTEENTH- AND nineteenth-century America, in rural
England well into the twentieth century, and for hundreds of
years in India, many potters split their year between pot making and
farming. During the inclement weather of winter, the potter made
pots. During the growing season, he fired his wares and tended to his
crops. The two tasks, both of the earth, both physically demanding,
both dependent upon luck; the vagaries of the kiln fire and the
vagaries of the weather are more similar than not, and fit nicely
together with the changing seasons. A potter can make some extra
cash raising a food crop, a farmer can make the long dark days of winter productive by sitting at his wheel.

Potter-farmers, needing to ease the workload of a hard life, came
up with ingenious solutions to the agricultural life. Ancient Egyp-

tians built little incubators of earthenware tiles and set them over heaps of fresh camel manure, which gave off heat as it decomposed, and warmed the eggs and hatchlings of the guinea hens they raised for eggs and meat. Egyptians also came up with a way to carry water to distant fields, even uphill from the Nile or irrigation canals, without exerting themselves too much. This invention was called a shaduf (sometimes shadoof)—a long pole mounted like a seesaw. On one end was a jar, on the other, a weight made of clay. The farmer would pull on a rope to lower the bucket into the water. When he released the rope, the counterweight of clay would lift the jug of water. Sometimes a series of shadufs were constructed, one after another so that the water could be carried across a field and uphill with ease. Similar devices are still used in India today.

For less arduous watering, jugs or jars would have been used from the earliest days of plant care. It was a simple matter to fill the water jar from the well and pour water on seedlings. By the Middle Ages, European potters were producing robust watering jugs. A pierced disk of clay covered the spout, forming a "rose" for sprinkling, which enabled the gardener to give the leaves a gentle rainlike spray of water. There was always at least one handle, but sometimes two were added, one on the side and one strapped across the top, because a jug of water is heavy. The jugs usually had a half-disk shield across the spout side of the mouth so that when the watering jug was tipped all the way, water did not gush out the top, and instead was forced down the spout and out the rose. Lightweight metal, galvanized tin or copper, has replaced clay as the material of choice for what we now call watering cans, though traditional clay watering jugs can still be found in Italy and France.

Crops are not the only watering concern of farmers.

Livestock needs water too.

Chickens, in fact, need an abundance of fresh water to drink, but if you fill a bowl or trough with enough to last the day, they are very likely to hop in, which is unsanitary, and occasionally they even drown. To tackle this agricultural challenge, English and American country potters made a bell-shaped dome of earthenware (and later of stoneware) with a knob or handle on the top, and a small round hole about an inch from the bottom edge. They also made a matching shallow dish, with a wider circumference than the dome. To use the two-piece "chick waterers," or poultry fountains, the farmer (or more likely the farmer's wife) would turn the dome upside down and fill it with fresh water. She would then set the dish over the base of the dome while it was still upside down, and flip the two over together and set them on the coop floor or in the barnyard. A vacuum formed in the top of the fountain so that the water in the dome would fill the dish to the level of the hole in the dome, but not spill out. As the chickens drank, more water was let out of the dome into the dish. This ingenious device kept the chickens refreshed and saved the farmer from having to spend the day refilling. The fountains were usually made large enough to hold a two- or three-day supply of water.

Another, probably earlier style was made in one piece. It was essentially a jug with a cup-shaped opening near the base. It was imperative that the lip of the cup be pulled out higher than the interior wall, so that when filled, like the two-piece fountains, a vacuum was formed. Sometimes these were made instead with a second half jug attached in the front with a cup-shaped opening. There is a poultry fountain in the collection of the Greensboro Historical Museum in Greensboro, North Carolina, that looks like a face. Poultry fountains were sometimes used to feed chickens buttermilk for extra nourishment.

Unlike the one piece fountains, the two-piece poultry fountains, though more bother for the potter to make, allowed several chickens to drink at once. A hen could gather with her chicks or several members of the backyard flock could sip together. And because it could be taken apart, the two-piece was easier to clean.

English potters also made chicken feeders, which were somewhat similar. They had a hopper, which supplied a small trough with feed. Dinner became a self-service affair in the chicken coop.

Alessandro Finzi, who has written an online manual for trainers and technicians who assist families in poor and undeveloped countries as they strive to gain food security, suggests that poultry fountains, or siphons as he calls them, can be hand molded and fired by even the nonpotter for use in the henhouse.[1] He suggests that they be kept out of the sun, especially in hot countries, because the water will heat up, though he also points out the cooling effect of evaporation from a pot that is made of earthenware.

Finzi, who is Italian, has other ideas for using pottery when keeping backyard chickens. He thinks a bowl, especially a removable one that sits on a pedestal, makes a good mobile nesting place for a hen to lay her eggs. He points out that ceramic is easily cleaned and cannot harbor pests the way wood can. He suggests storing eggs in layers interspersed with damp cloth and sand in a clay container, much like layering lasagna. To test for freshness, he explains that an egg should be placed in a bowl of water. If the egg remains horizontal submerged at the bottom of the bowl, it is fresh. As the egg ages, the wider end will gradually rise. An egg that is vertical is at the limit of edibility. One that floats is too old to eat. His last idea, which he shows in use in rural Africa, is a small mud-built chicken coop.

In the developed world, chick waterers are still in use, but they are now made of plastic, which is not very pretty and is less sanitary than stoneware. However, they are lightweight and inexpensive. Antique ceramic poultry fountains, like so many old-fashioned farm implements, have become collectibles. A particularly lovely salt-glazed chick waterer with deep blue cobalt brush strokes across the front, which was made by a potter in Pennsylvania in 1875, recently sold for over five thousand dollars at an auction. Only the most high-class chicken could afford to drink from it now, but it is too pretty to relegate to the barn anyway, and it can take its place next to works of art in the best-decorated living room without causing any raised eyebrows.

Many farmers kept rabbits, so potters in both Europe and North America made feeding dishes that fit their particular needs as well. Like chickens, rabbits are inclined to climb into their food and soil it, so potters devised small straight-sided dishes with a lip that curved inward to discourage such behavior. They also made drinking dishes with a simple straight lip. Both the feed and water dishes were made thick and heavy so the rabbits could not knock them over. And if you were a farmer sitting at your wheel in winter, making dishes for the rabbits in the hutch by the back door, why not make dishes for the family cat or dog?

Although it wasn't until fairly recent times that farmers and gardeners understood the relationship between bees and pollination and hence the importance of bees to crops, farmers have kept bees or "encouraged" them for their sweet honey for millennia. We know from extant paintings that Egyptians had hives as early as 2400 B.C.E. Archaeologists have discovered hives from 400 B.C.E. in southern Greece and from 250 B.C.E. in Spain. Romans wrote about hives as early as 146 B.C.E.

Hives could be made of old logs or baskets, but generally they were made of fired clay, though in some instances, they were constructed of unfired clay. In eastern Crete, where they are called *solines* (pipes) and are made of earthenware, they have remained unchanged from the late Turkish period to the present. As in much of the ancient world, these pottery beehives are set on their sides. Archaeologists have found numerous shards of Greco-Roman beehives and their extension rings and parts of lids on the island. In western Crete, upright hives are used. They too are usually made of terra cotta, though hollowed-out trunks or boards are sometimes used.[2] Bees were killed in order to harvest the honey. Like most of the hives, the fumigator was made of pottery.

Dairies, even small one-cow family operations, required a supply of milk pans, large bowls with flaring sides, almost like giant pie pans, to use for separating the milk from the cream. They also needed butter churns so that the cream could be beaten into butter. Until the nineteenth century, churns were made of wood. Then, in both the United States and in England, potters began making churns. These were usually tall, often two feet or so, and equipped with a lid with a hole in the center. Around the hole was a raised and flared cylinder of clay, which would catch the buttermilk splashes. The dasher was still made of wood and slipped through the hole. At the end of the dasher, there was either a wooden disk or cross that beat the cream. Sometimes the lid was made of wood rather than clay, as it would make less noise than a clay lid. Ceramic churns, though heavy, were more hygienic than wooden churns, which were absorptive.

At the beginning of the twentieth century, geared churns came into use. They were more usually made of glass, but potters made them of stoneware. They had a wheel gear on top, and an eggbeater-type contraption inside, and were advertised to churn butter in one minute.

Dairy farms also require a way to store enough food so that cows can sustain milk production through the winter. The ancient Greeks and Romans stored fermented animal feed in airtight underground pits called, in Greek, *siros*. Asians and Germanic people also stored grains and food in underground pits. By the third quarter of the nineteenth century, Europeans were experimenting with improvements. They built long stone storage facilities and moved them above ground. In the United States, farmers, following the advances in Europe (particularly France) and adding their own improvements, built tall, rectangular silos, of stone or wood, followed by round silos, generally of wood. At first the wooden silos were of balloon construction or sometimes of stacked wood, but by the close of the century they were of stave construction, in effect, towering barrels.

Twin silos made of salt-glazed tiles on the campus of the University of Connecticut.

In the first decade of the twentieth century, some silos were lined with brick or built of brick to guard against vermin and fire. Poured cement, cement staves, concrete, and salt-glazed tiles all came into use as building materials for silos at this time. In addition to silos, salt-glazed tiles were used for foundations and for milk houses. The glossy, deep-rust-colored tiles were durable, impervious to damage from acids leaked from the silage, easy to clean, and attractive. There were some problems with the mortar, which was not as waterproof as the tiles, but new mortars were found.

By the twenties and thirties, salt-glazed tile silos were common. However, after the Depression few of these silos were built, and by the sixties, steel silos, bright blue pillars in the sky, lined with fiberglass, dominated. Old sagging and leaning wooden silos are a common sight in the twenty-first-century countryside as farmers' children turn to other careers and, neglected, the silos decay. But many tile silos still stand sturdy and erect, even after their roofs have caved in. They are still an attractive feature of the rural landscape and some remain in active use eighty years after they were built.

Insects are often a problem on the farmstead, whatever agricultural product is produced. Eighteenth- and nineteenth-century potters in the American South made special ant traps for food safes and tables. The traps were shaped like squat cups with an interior space for the leg of the table or safe, and an outer space or ring, which was filled with water or, more drastically, kerosene. In England, special beetle traps were made. These were round and conical shaped, with a series of unglazed steps that led to a shiny-glazed well. To use the trap, the farmer would fill the well with a fragrant liquid, enticing the beetles to climb in and, like the hapless ants, drown. Roach traps also relied on luring and drowning their victims.

Potters offered two solutions to pesky outdoor insects such as mosquitoes. They made small portable furnaces of earthenware or heavily grogged stoneware during the first three decades of the twentieth century. These thick-walled, pail-shaped furnaces were equipped with a clay grate and a metal handle and were usually encased in metal. Charcoal was burned in them. These furnaces were often used on laundry day to heat water for the wash, to heat the iron, or

Salt-glazed silo adjoining a restored barn on the University of Connecticut campus.

even to warm the washroom. However, if you set them down outdoors and tossed damp leaves and green wood on top of the charcoal, the furnace turned into a smudge pot and kept insects at bay.

A more elegant solution to the insect problem was the birdhouse. Archaeologist Audrey Noël, author of *Colonial Gardener*, explains that "for centuries, wise gardeners have known that the best and cheapest way to control insects lies in the encouragement of their natural enemies, of which birds are the most responsive. In Virginia, where mosquitoes could sharply diminish the pleasures of colonial gardening, the martin was a welcome visitor. In order to encourage them to remain close to a garden, the provision of a suitable nesting place was essential, and for this pottery bottles were hung beneath the eaves of houses and outbuildings."[3]

Early potters throughout the South, in Texas, and in New England made simple birdhouses from the traditional jug shape. A circular opening was cut in the front or the whole jug might be turned on its side, and the mouth widened for entry. Some of these bottle or jug birdhouses came equipped with little clay roofs over the entry and holes where a wooden perch could be added. They all had accommodations for hanging.

Over time, designs changed and sometimes no longer resembled a jug at all. Early twentieth-century potters also began making terra-cotta birdbaths, especially in the American South where winters were not cold enough to freeze them. Birdbaths usually consisted of a column, wider at the base than at the top, with a removable shallow basin. In colder climes, the basin would be emptied and left upside down during the winter.

Clay birdfeeders also entered the potter's repertoire. These could be a simple, shallow bowl, suspended from a tree branch with string, or a more sophisticated construction, made in sections, with a dish for birdseed, a refillable hopper, and a roof. Today, ceramic birdhouses, birdbaths, and birdfeeders are all readily available to the home gardener.

Although farmers and gardeners appreciate the "integrated pest control" that birds offer, voraciously eating many of the hungry insects that devour fruits, vegetables, and the leaves of both decorative and edible plants, and they enjoy the benefits of the birdhouses and birdfeeders that assist them in attracting birds to combat these troubling insects, many birds are themselves considered pests. They eat the corn before it can be harvested, topple sunflowers just as the seed heads ripen, and strip berry bushes clean before you can say blueberry pie. In addition, four-legged mammals have a habit of helping themselves to whatever is ready to harvest.

Potters have not played a major agricultural role in combating these unwanted guests at the feast, but in India, they have come up with a unique form of scarecrow. Sometimes old cracked jars are placed on poles in the fields, but potters more often make vessels particularly for this purpose. They sculpt frightening faces onto the upside-down pots before firing them. These grotesque characters are quite similar in appearance to the face jugs of the American South. Placed near the outer edges of a field, they make quite a first impression. It is difficult to say how effective these ceramic scarecrows are, but they inspired me to make a faceless version with suspended chimes in hopes of shooing away the deer that invade my gardens.

"Scarecrow" by the author, inspired by scarecrows made by potters for farmers in India.

A modern pest control fad has been the introduction of "toad houses" into the garden. These can be plain red earthenware domes with a doorway cut in one side or elaborate terra-cotta faux castles complete with turrets, but the principle is the same. Toads like cool damp shelter, which the houses provide, and toads eat huge quantities of insects. Theoretically, they find the houses so inviting, they move into the garden and munch on the bugs.

Ming dynasty gardeners were not as interested in having birds and toads in the garden as they were in having fish. These were both a decorative element, like fishponds are today, and an early version of a fish farm: they ate their fish.

The imperial porcelain factory of the Ming dynasty was pressed into duty to make the fishbowls for the emperor's garden. This was not an easy assignment. Fishbowls became larger and larger until a typical bowl, flat bottomed, was thirty and one-half inches in diameter and about two feet deep. Bowls were decorated with dragons, waves, waterweeds, and fishes in blue and, sometimes, other colors. They might have an outward-turning rim or swelling belly. They were showy, impressive to behold, which was the intent. But they were a nightmare for the potters in the imperial factory. Pots of this size, particularly with such a wide base, were at the absolute limits that could be made and fired. Often, in fact, they failed in the kiln, splitting across the bottom. Frequently, the potters protested an imperial order for fishbowls, as the ever-larger renditions that were requested were so difficult to make.

In the years following the Ming dynasty, the empress ordered fishbowls that were three and a half feet in diameter and two and a half feet high, which the potters at the imperial factory complained could not be done. Nevertheless, they attempted the feat for four long and arduous years, but each beautifully wrought fishbowl that they made came out of the kiln cracked. Finally, the empress realized that she had asked the impossible and set aside her request.[4]

Porcelain garden seats, which were also popular in the Ming dynasty, were much easier to make, though they too were large. A glowing description of these porcelain seats appeared in a survey of art objects published during the Tianji (T'ien Ch'i) period (1621–27). It explained that "there are the beautiful barrel-shaped seats, some with openwork ground, the designs filled in with colours, gorgeous as cloud brocades; others with solid ground filled in with colours in engraved floral designs so beautiful and brilliant as to dazzle the eye: both sorts have a deep green (ch'ing) background. Others have a blue (lan) ground, filled in with designs in colours like ornaments carved in shih ch'ing (stoneblue)."[5]

These could not have been comfortable except for perching or sitting for a brief rest, perhaps an interlude of meditation, but they must have looked splendid amid the tree peonies and chrysanthemums. It is easy to imagine a European or American farmer, under much less refined circumstances, using an upturned crock for a moment of repose. Victorians loved the idea of ornately glazed garden seats and, if funds permitted, decorated their gardens with reproductions and imitations of Ming ware. One late-nineteenth-century German factory made a heavily decorated version complete with a "pillow" on top. The pillow, also ceramic, was designed so that rainwater would run off. Modern-day inspirations and reproductions can still be had, though as often as not, decorators keep them indoors.

Rhubarb pots, a more prosaic horticultural item than the Chinese-style garden seats, were also fashionable in the Victorian era. In 1815, the rhubarb bed in the Chelsea Physic Garden was trenched in the autumn. To everyone's surprise, the covered plants matured early. The following year, the gardeners covered the plants with pots and covered the pots with earth and, to their great pleasure, duplicated the previous year's success. The best method eventually worked out was to cover the dormant rhubarb in the fall with a tall pot. The pot was then buried under alternating layers of fresh manure and leaves that had been raked up from the yard. The leaves and manure gave off heat as they composted and by February they had completely turned to "black gold," which was cleared away for use on the vegetable beds. Inside the forcer, the rhubarb, tender and blanched white, was ready to eat. Similar treatments were given to sea kale and chicory with happy results. This process could only be repeated on a plant every three years or it would weaken. Soon tall, bottomless forcing pots, with lids that opened, were seen in every respectable Victorian kitchen garden.

Down the centuries, potters have been resourcefully serving farmers and gardeners with the products they make of clay, often through firsthand knowledge of the agricultural or horticultural tasks to be performed or problems solved. There is seemingly no limit to what can be made with clay for use in the garden or on the farm. British potters of the eighteenth and nineteenth centuries made cone-shaped earthenware molds so that sugar could be made and molded from a boiling solution of the canes grown in the Caribbean islands. Texas potters made similar molds for Mexican sugar manufacturers in the early twentieth century.[6] Potters have made tree rings for saplings to keep mice away and tree rings that irrigate with a slow drip. They've made plant markers, lanterns, wind chimes, and fountains for the garden. They've made rain drums and bells.

Scientists add heat-treated bentonite clay to herbicides and pesticides to act as a carrier during application. They add it to animal feed as a supplement. It also aids in pelletizing commercial feed pellets. It helps soy meal feed flow better. Soil scientists use bentonite as an ion exchanger to improve soil.

But it was the invention of pottery, and the storage jar, that led to the invention of agriculture itself as we saw in chapter I; it was clay jars that made seed storage viable.

And it was the flowerpot that most impacted horticulture. We don't know when the first flowerpots were made. The oldest known depiction of a flowerpot, from prior to 2000 B.C.E., is on the face of the altar of the temple of Hagar Qim in Malta.[7] The pot has fairly straight sides and some sort of embellishment on the rim. The plant grows straight up and does not have flowers. The plant and pot are so contemporary looking that they could be a foliage plant for sale at a local nursery.

We know from illustrations on tomb walls that the Egyptians used planting pots. Queen Hatshepsut (1495–1475 B.C.E.) sent a party to the land of Punt to trade some of her country's goods for a collection of aromatic myrrh trees, dug up with their roots intact, which she planted in pots at her palace. Three and a half centuries later, Ramses III (1198–1166 B.C.E.), took up gardening in a most serious way. He funded 514 semipublic temple gardens. We know from wall paintings that he grew small shrubs in decorative earthenware pots and that the gardens were formal and geometric. There were arbors, "garden rooms," trees, walls, ponds, and, set out as focal points, generously sized terra-cotta garden pots.

The Egyptian style of gardening was taken up by the Greeks and Romans and by inhabitants of the Near East. It was in Greece that the seasonal cult of the Adonis gardens began and from there spread to Rome and the eastern Mediterranean lands.

Adonis was the breathtakingly handsome prince of Cyprus. Aphrodite took one look at his well-muscled body, his high cheekbones, his clear eyes, and she was totally enamored of him. Although she had other men in her life, it was Adonis that she wanted. Unfortunately, Persephone, the queen of the Underworld, was also besotted with the gorgeous Adonis. She was wildly jealous of her rival Aphrodite and concocted a plan to make Adonis her own. She told Aphrodite's lover, Ares, about his girlfriend's crush on the pretty boy Adonis, knowing that she could provoke his hot temper. She did. Enraged, he turned himself into a fierce wild boar and while the unsuspecting Adonis was in the forest hunting, Ares the angry and jealous boar attacked him. Wounded, Adonis bled on the earth and in the spots where his blood dropped, anemones began to grow. Adonis died and descended to Hades, smack into the open arms of Persephone. Aphrodite was furious; she ranted and raved and complained to Zeus. Finally, he ruled that Aphrodite and Persephone would share Adonis, with each having him for six months of the year. Adonis's descent to Hades came to symbolize the onset of winter, and his rise, the renewal of spring.

Versions of the myth go back to Mesopotamia and appear throughout the ancient world.

In Greece, to commemorate the seasonal change, women would set a statue of Adonis on their rooftops toward the end of summer. They would ring the image with flowerpots, which were usually amphorae that were broken. We know this from pictures of the celebration on Greek vases. The pots were filled with soil and planted with lettuce and fennel seeds. The seeds sprouted but the women did not water them so the seedlings shriveled and died. This symbolized the change from summer to winter.

Roman women made Adonis gardens on their rooftops too. By the beginning of the Common Era, it had become a huge national holiday with lavish celebrations and festivals.

But Romans liked the looks of flowerpots on their rooftops too much to limit themselves to festival time. They lined their balconies with pots of herbs and flowers, set pots on their rooftops, and filled their narrow window ledges with terra-cotta pots of plants. They placed pots and urns in their courtyards, around their fountains, and as focal points in their porticoes. We know about Romans' use of garden pots from their frescoes, from excavations, particularly at Pompeii and Herculaneum, and from the writings of Cato, Pliny the Younger, Pliny the Elder, Theophratus, and others.

Roman flowerpots had a hole in the base and three root holes on the sides. They were used to transport exotic plants great distances, to plant trees and shrubs in planting pits dug in the ground or rock, and for propagation.

Roman horticulturalists developed several methods of propagation, each requiring a growing pot. The first way was to bend the stem of a healthy tree or shrub to the ground, insert it into a soil-filled pot, and bury or cover the whole business. Once rooting took place, the new plant could be severed from the mother plant and then planted out, pot and all. Another method was to suspend a soil-filled pot in a tree and insert a growing tip into it (air layering). And the third, of course, was to plant seeds in the pot.

With the many advantages that flowerpots offered, their use by the Romans was widespread. Plants could be grown in vast numbers in nurseries. Soil could be amended. Moisture could be retained. And perhaps most important, plants could be conveniently transported.

At least some of the time, Roman gardeners cracked their pots before setting them in the ground, but left the pieces around the root ball. Archaeologists have found evidence of pots that were presmashed, pots that seemingly broke after planting when the roots bulged, and intact pots. Because the pots usually had a smaller mouth than belly, once a plant was well established, it would be difficult to remove.

Chinese gardeners, though heavily reliant on architecture, used curves and elements of surprise in their gardens rather than the geometry of the Egyptians, Greeks, and Romans. At least as early as the Tang dynasty, and probably earlier, they were making pots for specific use as flowerpots. They placed these glazed pots near a path or a seat or in a corner of their garden. They were round, square, rectangular, and octagonal and usually heavily decorated. Some had scroll feet.

The Chinese loved flowers. They made pots with narrow necks to show off a branch of cherry blossoms. They grew the sacred lotus, chrysanthemums, and peonies and might pluck a bloom and hold it up to study. They also forced bulbs and made pots for that purpose. During the Song dynasty, garden historian Anthony Huxley writes, "Itinerant flower sellers carried such containers in flat panniers swinging from a yoke on the shoulder; the peddler would take away containers whose plants had finished flowering and nurse them into the next season's bloom."[8]

Bonsai, which we think of as Japanese because it has been so expertly practiced in that nation for so long, originated in China and actually means "grown in a tray." Huxley tells us, "The choice of an appropriate container was supremely important and many were works of art in their own right, usually rectangular but sometimes oval or with rounded corners. These containers would be set aside in special places, the main collection was housed under a screen of lathing to diminish the sun's heat while allowing maximum air circulation, and specially chosen specimens were set out on stone benches or on the edge of a pool."[9]

Bonsai is popular today in the West as well as the East. The proper container remains almost as important as the plant and though commercially made pots are available, they are always ceramic. There are studio potters who specialize in making them, usually one of a kind.

The great Islamic gardens also included flowerpots but they contained plants valued for their scent rather than for their showy inflorescences. These gardens, designed with powerful symmetry, were enclosed and offered shade from the pounding heat of the daytime sun. They were rich with symbolism and featured four intersecting paths based on the confluence of earth, air, fire, and water.

Pots were placed where the paths met and generally held only one plant or one type of plant.

Even during the Dark Ages, though much knowledge was lost, at least to Europeans, flowerpots continued to be made and used, though they were not especially grand. The formal lines of Roman and Islamic gardens remained the ideal. In the fifteenth century, as Italy emerged from the dark years, the architect Leon Barrista Alberti promulgated what he called the Italian Style, though he drew heavily on the works of Pliny. He advocated the use of flowerpots planted with herbs and violets.

A Florentine, Luca Della Robbia (c. 1400–1482 C.E.), a goldsmith and stone carver, turned to ceramics because he knew that he could make sculptures in clay more quickly than in marble. He wanted results. He hired his relatives and established a family workshop to produce his designs. He was not in business long before he was known for his terra-cotta bas-relief plaques and earthenware figures. The Della Robbias also made large fancy flowerpots of earthenware finished with raised swags and leaves, by hand pressing soft clay into capacious fired clay molds. Wealthy owners of large manor estates ordered the pots for their expansive gardens. Knockoffs are common today.

Since the time of the Egyptians, with greatly increased activity during the Roman era, plants were dug up and transported many miles from their place of origin to be replanted in distant lands. After Rome, this activity abated somewhat, but the seventeenth century saw the rise of the plant hunters, explorers who deliberately set out to find new botanical specimens. With this, farming and gardening saw the introduction of many new plants and interest was excited.

In England in particular, a whole industry of experts rose up to offer gardening advice in books and magazines, and often to sell the plants and seeds. John Evelyn, one of the experts, wrote in *Elysium Britannicum*, published in 1654, "[Pots] may be employed for the sowing, setting in and preserving of the choicest of flowers . . . and therefore to be made of various sizes, depths and diameters, frequently and commodiously enough moulded of common potters earth, but always pierced at the bottom, for the passage of superfluous showers, which otherwise would overwash rot and starve the roots contained."[10]

Nurseries required large numbers of flowerpots in which to offer their plants. As the bustling factory potteries of the Industrial Revolution took over the production of tableware and families no longer had to produce all of their own food because trains could economically transport meat, produce, and dairy products across country, rural potteries turned more and more to the production of these much-in-demand flowerpots. Though many country potteries did close, hand-thrown flowerpots were made in huge quantities until the middle of the nineteenth century. This was true in the United States as well, especially in the South.

Pot sizes and shapes were standardized. Now the base was narrower than the top and the sides were straight: graduated pots that could be stacked. This design, which is still the standard flowerpot shape, eliminated the problem that the Romans had, and made repotting an easy task. If the top is wider than the bottom, then the root ball can easily be slipped out. There were also tall, narrow pots (called long toms) and shallow pots (called seed pans), thimbles, saucers to put under pots, orchid pots with multiple perforations on the walls, and forcing pots for bulbs.

In 1769, an independent and entrepreneurial English woman, Eleanor Coade (1733–1821) went into business to support herself. She set up a factory to manufacture flowerpots, and was soon making remarkable pots that could withstand British winters. After an employee, Daniel Pincot, who may have actually been the one to create the secret clay formula for what became known as Coade stone, began representing himself as Coade's partner to clients, Coade fired him in a huff and hired the sculptor John Bacon. Bacon used the secret recipe to reproduce classical statues and urns. Business soared. Anyone who had a bit of money wanted something made of Coade stone. Coade stone urns, statuary, and building ornaments began to appear on British estates. The formula, which included clay, sand, flint, crushed glass, and grog, was lost when the factory closed, but in 1985, a British stone carver re-created it and it is now available again.

During the Victorian era, plants came indoors. Ferns, aspidistra, palm trees, and agaves filled the front parlors of every respectable family. The rich built conservatories and glass houses to overwinter their prized orange trees. Houseplants were considered beautiful and healthful. It was good for you to be surrounded by greenery. Virtually every household needed flowerpots. And jardinieres, glazed outer pots, were useful too, to protect the finish on fine furniture from the dampness of the earthenware plant pot.

Ships, laden with exotic plants, plied the seas with their botanical cargoes. Merchants grew wealthy. Captain William Bligh, of *Mutiny on the Bounty* fame, who was hauling breadfruit plants from Tahiti to the West Indies, wrote in his diary in 1878, "The great cabin was appropriated for the preservation of the plants ... It had two large sky lights, and on each side three scuttles for air, and was filled with a false floor cut full of holes to contain the garden pots, in which the plants were brought home. The deck was covered with lead, and at the foremost corners of the cabin were fixed pipes to carry off the water that drained from the plants into tubs placed below to save it for future use." [11]

Bedding plants, which also hailed from distant lands, were all the rage for outdoor Victorian planting schemes. They required thousands of pots for starting.

But soon, even the production of plain earthenware flowerpots was mechanized. The remaining rural potters in the American South made strawberry pots, hanging planters, and pots with frilled rims. More British potters closed up shop, but a few continued to find customers for their traditional wares throughout the twentieth century. Then, between 1961 and 1963, the nursery industry switched to plastic pots. Within a few short years, nearly every garden center and nursery sold virtually all of their offerings in plastic. You could still purchase a clay flowerpot, but plants were not sold in them.

Ironically, a decade later, the back-to-the-land movement was sweeping the United States, craft fairs were sprouting up, and ambitious young potters were offering hand-thrown planters with macramé hangers. This author published a little booklet for people with no special training who wanted to make their own clay flowerpots rather than tolerate plastic.

Of course, flowerpots are not the only ceramic ware associated with flowers. Vases have been made for millennia, though it is not always possible to know if indeed they were used for cut flowers; the Chinese put flowers *and* branches in their vases. During the height of tulipmania in Holland, potters made multinecked vases to show off each stem. The Greeks and Romans are thought to have put flowers in their vases.

More than flowerpots, vases have tended to reflect the fashion of the moment. In the last two hundred years we have seen Art Nouveau vases with matte glazes, jazzy Art Deco vases, art pottery vases influenced by the Arts and Crafts movement, and an array of mass-produced vases in styles imitative of almost every era in history. Ikebana artists, trained in Japanese flower arranging, take particular care when choosing a vase, often a shallow bowl with a ring to hold a single stem. Pansy rings are made to hold a "wreath" of cut pansies or violets. Clay frogs, perforated disks inserted into the bottoms of vases, or clay "pebbles" assist the flower arranger and keep the flower stems from flopping gracelessly.

■

IN THE TWENTY-FIRST century most of the farm and gardening implements mentioned in this chapter are no longer made of clay. However, bentonite continues to play a major role in farming. There is a resurgence of interest in gardening, which has led to full-color ads in the backs of specialty magazines offering enormous oil jars from Cyprus and Turkey, traditional thrown flowerpots from England and Tuscany, and antiques from the world over. Discount stores offer large Vietnamese flowerpots with appealing blue and dark brown glazes. They are made from clay so rough most of the pots have blowout scars and sell for prices so low one wonders how it is possible to ship them around the world let alone make them. You can still purchase a plain clay flowerpot at most garden centers. Every year publishers announce several new books on "container gardening." It does not look like clay and plants will be parted anytime soon.

ELECTRICITY, TRANSPORTATION, AND ROCKET SCIENCE

Thou didst create the night, but I made the lamp.
Thou didst create clay, but I made the cup.
Thou didst create the deserts, mountains and forests,
I produced the orchards, gardens and groves.
It is I who made the glass out of stone,
And it is I who turn poison into antidote.

—SIR IQBAL MUHAMMAD

FORGET BENJAMIN FRANKLIN and his famous kite. Forget Thomas Edison and his lightbulb. Forget Alessandro Volta and his battery. The "Baghdad battery," capable of generating between 1.5 and 2 volts, was made around 200 B.C.E., two millennia before these men were even born!

It was discovered in 1938 by German archaeologist Wilhelm König. It is a small 5½-inch-tall earthenware jar, about three inches across. Enclosed within is a copper sheet rolled into a tube. At the base of the tube is a copper disk, held in place with asphalt. The top of the jar is sealed with asphalt. An iron rod is inserted through the asphalt seal and suspended inside the copper tube, straight so that it does not touch the copper. If the jar is filled with an acidic liquid, wine for instance, and the two metal "terminals" are connected, then the liquid becomes an electrolyte and permits the flow of electrons from the copper tube to the iron rod: yes, electricity. Scientists have conducted experiments with the jar, and it most certainly does work as a battery.

Since 1938, many other earthenware batteries have been found in Iraq. What were people *doing* with batteries two thousand years ago? Copper vases coated with silver dating from the same time period have been found in Mesopotamia and Egypt, though how these vases were made has not been exactly determined. Scientists have hypothesized, however, that the ancients must have had some knowledge of—and a method for—electroplating. Electroplating is the process of using an electric current to coat something with metal. If several Baghdad batteries were linked together, they would have generated enough voltage for the electroplating process.

Another thought is that the batteries were used for some sort of shock treatment. We know that the Greeks and Romans used electric eels for pain treatment (on the theory that a worse pain makes you forget your complaints?), so perhaps the jars were used in this manner. Unfortunately, we may never know for sure what the batteries were used for, especially since the concept was lost until Volta made his battery in about 1800.

Ceramics have been important to the development of technologies since the first potter fired the little figure she had modeled in mud. Kilns, wheels, glazes, mold making, the construction of buildings, chemistry, and other milestones in the progress of civilization were dependent on clay. These developments led to advances in other fields.

Clay and the technical skills acquired in producing ceramics played a critical role in the metallurgic arts and ushered in the Bronze and Iron Ages. Metallurgy requires several processes: mining (or in the case of gold, panning), extracting, smelting, alloying, annealing, and forming. Clay was sometimes used to face the walls of mine shafts, often in conjunction with wood reinforcements. However, ceramic technologies and clay itself were more critical to other metallurgic processes.

Copper, which turns a very noticeable turquoise or teal color when exposed to the elements, was at first easily gathered. Nuggets lay near riverbanks. It glistened in the faces of rock walls. By 4500 B.C.E. the Chaldeans were prospecting for it and, because it is a malleable material, they were able to beat it into useful shapes. Somehow it was learned that if copper is heated first, it is much stronger when hammered. This heating process, possibly inspired by the heat transformations of ceramics, is called annealing or tempering. Now the bright orange metal was used to make knives, weapons, tools, and ornaments.

But the supply of readily available surface copper was soon largely depleted, and consequently early metalsmiths had to turn to minerals and ores, from which the metal had to be extracted. A mineral is a substance such as cuprite, which is copper combined with oxygen. An ore is a "mineral or aggregate of minerals from which a valuable constituent, especially a metal, can be profitably mined or extracted."[1]

To extract metal from ore, one needs heat. Again, we see the influence of potters, who had thousands of years of experience working with fire. Primitive metalworkers layered chunks of ore with fuel, especially charcoal, which produces more heat than does wood. After the fire died down, a lump of metal covered with slag, or waste, could be pulled from the ashes. The slag shattered easily and, once removed, left a chunk of nearly pure copper.

A more sophisticated method of extraction was to use a clay-lined pit furnace for smelting. Here, the fire could be better controlled and brought to a higher temperature.

Gold and copper were the earliest metals mined and worked. Gold was used for royal decorations: copper was the more practical metal. Agatharchides, a Greek geographer and historian, described the process of gold extraction in one of his books later quoted by Diodorus: "The workers place the crude gold in a clay vessel, and add a mass of lead, a little salt and tin, and barley husks. Then it is closed with a tight fitting lid, sealed with lute, and heated for five days and nights in a furnace. After a suitable interval for cooling, nothing is found of the other materials in the vessel, but pure gold."[2]

Clay crucibles, containers made of clay, were also critical to the forming processes of metallurgy. Once the metal was melted in a crucible, it could be poured into molds made of clay and cooled to form an object, or poured in sheets and hammered into shape. Crucibles were also used to make alloys—an alloy is a mixture of metals. Bronze is an alloy of copper and tin, usually 90 percent copper and 10 percent tin. It was the first alloy made and was tougher than copper.

The invention of metallurgy ushered in the Bronze Age, followed by the Iron Age (some archaeologists include a Chalcolithic or Copper Age between the end of the Neolithic era and the beginning of the Bronze era). Metals were used to make tools, pots, chalices, plow blades, arms, jewelry, and decorative objects. The advantage of metal, compared with clay, was that it could not be shattered like pottery. It heated up quickly for cooking, it was durable, and it was often beautiful.

Metallurgy spread from Mesopotamia to the Middle East, Anatolia, Asia Minor, and throughout Europe. Metal arts were highly developed in China and Asia and South America. As early as seven thousand years ago, copper was being worked in North America.

Though ceramics preceded metallurgy by millennia, once metal crafts were mastered, the two technologies flourished in tandem through the centuries. Sometimes the shapes developed in the fabrication of metal objects influenced the designs of pottery. Shiny gold and silver cups and bowls were often seen as status symbols by the royalty in cultures as diverse as the Aztecs of Mexico and the Egyptians of the Nile. Pottery, however, remained crucial to everyday life.

In Islamic countries, metal tableware was forbidden. This led to the development of exquisite lustreware.

Islam was founded early in the seventh century C.E. By the eighth century, a unified Islamic culture had spread to Persia, Arabia, Egypt, northern Africa, Spain, and the borders of India and China. Here, mathematics, literature, architecture, astronomy, ceramics, and the arts burgeoned. Trade flourished and merchants traveled great distances to conduct business.

Islamic potters, like their counterparts in Europe, were dazzled by the beauty of Chinese porcelain but could not reproduce it themselves. However, as we have seen, it was they who introduced and exported cobalt to China, where it was highly valued and used with grace.

Islamic potters used cobalt on creamy white tin-glazed ware to great effect. Images of people and animals were considered idolatrous and were forbidden, but as is often the case, restrictions led to creative solutions. Working within the rules of their religious leaders, Islamic potters covered their wares with curvaceous calligraphy, delicate flowers, leafy vines, and intricate geometric patterns. Their pots did not suffer from a lack of what was forbidden, but instead exemplified artistic genius.

Lustreware was an innovative answer to the stricture against metal tableware. Islamic potters discovered that if they painted designs on their already fired pots using powdered gold or silver or copper (metal oxides) mixed with a bit of water and perhaps clay, and then refired these pots at low temperatures in a reducing (smoky) kiln, the designs emerged from the kiln with a soft metallic sheen.

Lustreware is more understated than brightly gilt enamels or shiny metal pots, and has a rich and subtle complexity of tones that can only be achieved with skillful firing. This was not easy; the temperatures and the atmosphere in the kiln had to be just right or the pot would come out looking dull and dingy. Special muffle kilns,

Lustreware pitcher from Iran, circa twelfth to thirteenth century.

kilns within kilns, or kilns with enormous inner saggars were built to protect the pots from the direct flames yet allow the reducing atmosphere to permeate. Even with years of experience and proficiency, a potter making lustreware needed more than the usual dose of luck that all potters require.

The invention of glassmaking was also dependent upon ceramics and in fact, in scientific circles, glass is considered a ceramic. The first glasses, as we have seen, were glazes. The Egyptians discovered that by mixing ashes (potassium), ground-up sand (silica), and natron (salt from dried lake beds), they could give their pots a shiny coating. What they were doing was "fluxing" the silica. They learned by accident or through experiment that if they took a bowl of this glaze, especially one that had more flux and less silica, from the kiln while still molten, it could be poured into a clay mold and then cooled to form an object. Glass was equated with gems and was as highly prized.

Glassmaking progressed to what is called core-formed glass. Here, dung and soft clay are mixed and shaped into a solid vase or amphora shape, a rod is inserted into the center for handling, and then the clay-dung mixture is allowed to air-dry. This core is then coated with molten glass.

There are several theories as to how the Egyptians accomplished coating the core. One thought is that they used a rod, probably made of fired clay or metal, to pull hot strings of glass from a crucible and then, with great speed and dexterity, wound the hot strings around the core. The strings were then smoothed together with a second heat treatment. Another thought is that the core was rolled in crushed glass and heated. A third theory is that the core was dipped into a crucible of melted glass and coated.

When the object had cooled, the clay and dung core were scraped out, though some residue remained. Egyptians were making glass as early as 3500 B.C.E. However the objects were shaped, they were colored by adding oxides such as copper.

Interestingly, unlike pottery, the making of glass and the forming of glass into objects usually occurred as separate operations distances apart. The raw materials for glass were mixed and melted and then transported as rods or chunks for use by artisans.

Rods of glass were convenient to use with clay molds. Sometimes the rods were softened and curled into snail-like disks, which were then placed over a clay mold and heated until they fused. Other artisans coiled the softened glass rods into spirals around the interior of the bowl form before heating. Or molten glass could be poured into an incised clay mold similar to the press molds used in pottery making.

Beads and necklaces of bone, seeds, clay, shells, bits of wood, and metal have been made since the depths of prehistory. Glass beads, which could be made by heating and slicing glass rods, however, were smooth on the skin, lightweight, and resembled jewels. They became vehicles of trade as well as body adornment and status.

Glass differs from pottery, in that there are no pores. Pottery contains many separate particles which are tightly bonded together, but which have small spaces or pores in between them. Glass has no pores or spaces between the particles, and the particles have no boundaries. This is because the heat of the fire has melted the particles into a uniform liquid that is then cooled too quickly for particles to re-form. Glass is a liquid. Because there are no (or almost no) pores or spaces between particles, glass does not stop light, but instead, transmits it, letting it pass through. Glass then, is transparent.

Glass was made for three thousand years before glassblowing was invented, probably along the Syrian coast around 50 C.E. Glassmakers discovered that if they used a hollow rod to dip a blob of molten glass from the furnace or crucible, and if they then blew into the tube, they could create bubbles of glass. Just as the potter's wheel enabled potters to produce quantities of pots quickly, glassblowing allowed glassblowers to produce hollow glass vessels with speed and economy. Vessels could be free formed or blown into molds.

Romans exploited this new technology. Hollow glass vessels, the product of the glassblower's art, soon appeared throughout the empire. Glass was no longer only for emperors, kings, and pharaohs, but entered everyday life. Since the Romans, glass has been made into lenses for eyeglasses, microscopes, and telescopes, literally changing the way we see the world. It has been made into windows, dishes, "glasses" from which to drink, and a host of other necessities of daily life.

Both metallurgy and glassmaking, key steps forward for civilization, were outgrowths of the development of pottery. Each of the technologies has led to further developments.

Metal conducts electricity, but both glass and pottery act as insulators. From the earliest days of domestic electricity, clay insulators, often glazed a shiny dark brown with Albany slip, were used on telephone poles and for household wiring. In the United States, as the country was electrified, factories were devoted exclusively to the manufacture of these insulators, though occasionally small shops, especially in rural areas, did produce a few. Later, glass insulators were also used on poles. During the macramé craze of the 1970s, craftspeople collected old insulators that they found discarded beneath power lines and turned them into "hanging pots" for plants.

The main benefit to families when electric wires reached their towns was electric lighting. Now day could reliably be extended into night. Clay was important to ancient efforts to illuminate the dark. Simple earthenware lamps—open-topped dishes, often with a spout to hold the wick—were used to burn olive oil or animal fat in Palestine and Mesopotamia. Oil lamps were an important product of Roman potteries and appeared throughout the empire. Colonial potters made earthenware grease lamps and occasionally candlesticks. In India, earthenware oil lamps are lit in many regions for festivals. Victorian-era kerosene lamps were usually made of glass. Today, many lamp bases are pottery.

Clay insulators were also used in the first internal combustion engines. Clay, usually porcelain, was used as the insulating portion of spark plugs. In fact, the first spark plugs were thrown on the potter's wheel!

To make a spark plug, a metal pin was inserted into a thrown porcelain tube. A metal case surrounded the lower half. At the base of the metal case was a hook bent so that there was a gap between it and the base of the spark plug. When the car was started up, a spark jumped across the air gap, igniting the air and fuel and causing an explosion. The energy from the explosion moved the pistons, which then powered the engine.

Unfortunately, temperatures could reach as high as 4,000°F

(2,000°C), followed by rapid cooling. Porcelain is a good insulator and durable, but high temperature bursts and high voltages put a severe strain on the material. Accounts of early automobile trips are rife with breakdowns and stops for repair. Spark plugs rarely lasted for longer than fifty miles, so travelers carried spare plugs with them and made sure they knew how to replace them themselves.

In 1927, the president of General Motors hired custom designer Harley Earl (1893–1969) to head up the company's Art and Color Section. Earl, who had customized cars for movie stars while working in his father's design shop, was assigned the task of adding a bit of flare to new cars, thus setting the GM cars apart from others. Although he was to work within existing design parameters set by the company, within a few years he had revolutionized the automobile design process and the designs of cars themselves.

Earl instituted two major changes: he began drawing two-dimensional sketches for proposed new cars and he built clay models, a process he had begun in his father's shop in 1914. Prior to his tenure, models were built of wood and metal, a clumsy process at best. During his thirty-one-year career, Earl designed many cars, including the revolutionary 1953 Corvette. He came up with the idea of adding tail fins to cars and influenced and mentored many other automobile designers who adopted his methods. Later in life, he appeared in television commercials for new cars.

At first Earl's modeling clay was pure California mud. Today's modeling clay is made of a mixture of clay, wax, and sulfur additives. It is sensitive to ambient temperature and can be painted or shellacked. Designs are created on computers, but looking at, walking around, and touching a full-size or a scaled clay model is still the best way for designers, engineers, and executives to understand how a car will look. Computer-assisted milling machines do the initial rough modeling, after which modelers do the fine handwork. Modelers can quickly make any changes that management might want to see on a car. In fact they can usually make a change more quickly than the computer designers.

As the car changed from a novelty to a serious mode of transportation and the automobile replaced the horse and wagon, drivers lost patience with the need to constantly replace their spark plugs. Beginning in 1903, manufacturers began intensively researching a way to make the porcelain stronger. They discovered that by increasing the alumina content of the clay body, they could increase its strength. Gradually, the clay content was decreased and the alumina was increased. The body lost its plastic qualities and was no longer workable on the wheel. Instead, the powdered ingredients were shaped dry, under high pressure. Also, as the alumina content increased, the firing temperatures required for reaching maturity increased. Porcelain matures at 2,382° F (1,300° C), but with the body changed to 80 percent alumina, a temperature of 2,600° F (1,427° C) is required. With 95 percent alumina, the required temperature climbs to 2,950° F (1,621° C). [3]

Potters have been altering clay bodies almost as long as they have been using clay to make things; mix the slippery plastic clay from the riverbank with the stiffer clay from the bluff to the east, add a handful of sand, and maybe a bit of straw or dung, let the heap age, and then wedge it thoroughly before making cooking pots. However, experiments with increasing the alumina content (remember, clay is alumina, silica, and water) opened up vast new possibilities. High alumina bodies were shock and heat resistant, impervious to corrosion, and long lasting.

David W. Richerson, writing for the American Ceramic Society, explains, "Alumina was the pioneer of advanced ceramics, the model to be emulated for numerous other advanced ceramic materials that followed. The lessons ceramists had learned from creating products with alumina ceramics were applied to engineering all of the later ceramic compositions. Whole families of ceramics were invented: magnetic ceramics; piezoelectric ceramics; special optical ceramics. 'Designer ceramics' soon began to fill every demand of electronics, communications, medicine, transportation, aerospace, power generation, and pollution control." [4]

Although alumina is a key component of clay, modern ceramics is no longer, strictly speaking, mud. Modern ceramics is more broadly defined as "the art and science of making and using solid articles which have as their essential component, and are composed in large part of, inorganic, nonmetallic materials."[5] Advanced ceramics, as it is called, includes such materials as zirconium oxide, titanium carbide, silicon carbide, and boron carbide. Still, much of modern ceramics focuses on two of the key chemicals in clay: alumina, as we have seen, and silicon dioxide, or silica. Glass is fluxed silica. The tiles on the space shuttle are made of silica that has been engineered into a lightweight, heat-resistant, highly insulating material to protect the aluminum skin of the shuttle body. Ironically, the development of these tiles to protect the shuttle from the intense heat of ascent into space and reentry into earth's atmosphere led to the use of ceramic fiber blankets for modern factory kilns and many studio kilns too. The blankets are more energy efficient than bricks, lightweight, and cool quickly. I have a layer of this ceramic fiber cemented to the inside of my kiln door.

The electronics industry is the largest consumer of modern ceramics. Ceramics are a component of CD players, the electronic ignitions in cars, transistor radios, televisions, and satellite communications systems. High-silica ceramics are used for body armor to protect our military and police forces. Advanced ceramic knives, made of zirconia and toughened alumina, are superior to metal knives. They hold their edge indefinitely, and don't rust or stain.

Meanwhile, new uses are being found for traditional clay. An Italian acoustic designer, Francesco Pellisari, "on a visit to Umbria, the heart of Italy's ceramic industry, caught sight of a discarded ceramic cone. Pellisari's tests found that the rigidity of ceramics could be applied to speakers: the rigidity enhances the frequency of the sounds."[6] The manufactured speakers that resulted from this discovery are made of high-fired earthenware which is bisque fired, glaze fired, and then fired a third time with a coat of platinum. Not only did Pellisari's work enhance the sound quality of speakers, a dramatic new look was now possible, instead of the ubiquitous boxy speaker.

Thirty thousand years ago, our ancestors discovered that clay could be transformed by fire. This led, through the centuries, to many other discoveries and technologies. Today, ceramic engineers use highly technical equipment, formulas, calculations, science, laboratories, tests, and research to extend that ancient discovery into a future with as yet unimagined possibilities. Invisibly, silently (well, not so silently when it comes to alarm clocks!), ceramics are a key component of twenty-first-century daily life.

10

TO YOUR HEALTH!

I still believe that . . . mud will give you a perfect complexion.

—ZELDA FITZGERALD,
Save Me the Waltz

IN 1774 A Parisian dentist and surgeon, Nicholas Dubois de Chémant, made the world's first porcelain false teeth. Until then, false teeth were made of wood, bone, horn, or even teeth extracted from corpses. Installed in the patient's mouth, these organic materials would absorb stains, putrefy, and give off a dreadful stink. George Washington's famous wooden teeth would have made him a man with very bad breath. De Chémant was excited by the impermeability of the porcelain pottery that was being imported and manufactured in England and on the Continent. He realized that the impermeability of this "new" material, porcelain, would overcome the drawbacks of existing false teeth. Dubois de Chémant was right; his porcelain teeth were a vast improvement and a remarkable achievement.

However, they were not perfect.

They tended to become dislodged right in the middle of a meal.

Thirty-four years later an Italian dentist, Giuseppangelo Fonzi, inserted platinum rods into his porcelain false teeth, which he manufactured as single teeth rather than sets. His teeth were held in place much more reliably than Dubois de Chémant's.

Nevertheless, early porcelain false teeth were, in reality, pottery teeth. They were healthier and more attractive to use than their predecessors, but if the wearer bit down hard on a peach pit or bone, they could shatter. In the second half of the twentieth century, dental ceramics began to improve rapidly. Today, modern ceramics are used for "invisible" braces, false teeth, bridges, and caps. Specially formulated, these ceramics are nearly indestructible.

In the past four decades or so, the field of bioceramics has seen a burst of activity; intense research has resulted in lifesaving advances affecting all parts of the human body. Modern ceramics are used for bone repair, prosthetics, joint implants, and middle-ear implants. Ceramics are also used in diagnostic equipment, such as ultrasound imaging equipment, and in treatments such as dialysis and surgery. Ceramics are used in heart surgery, in pacemakers, and in respirators. If medical breakthroughs continue in this direction, we can say that the real "bionic man" will be made, if not exactly of mud, of ceramic material.

Clay's role in health care is divided between internal use, in the body itself, and a historic array of external accoutrements to treatment and research. It has played a role in practices shrouded in mystery and in misunderstandings about what is good for us. Yet it is key to twenty-first-century medicine.

Porcelain labware is sturdy, sanitary, and inexpensive. Porcelain mortars and pestles of various sizes: bowl-shaped evaporators with lips; Gooch crucibles, with tall sides and perforations; Büchner funnels, also with perforations; a wide range of crucibles and crucible covers; what scientists call casseroles (please, no baking macaroni!); and assorted spatulas and "plates" can be found on the lab benches of many research facilities. They come glazed, unglazed, and partially glazed. With their no-nonsense design and pure white color, they possess a forthright beauty, but they are meant for scientific precision. Porcelain labware is an important tool for chemists and pharmacists working to develop new drugs.

Pottery has had a long association with pharmaceuticals. Drug or apothecary jars were made and used in Persia, Syria, and Egypt as early as 1000 C.E. By the fourteenth century, they had found their way into Spain and Sicily and then to Italy, where they were called albarello. These jars are always tall and cylindrical in shape, with a gently curved waist and a short neck. The slim waists made the jars easy to grasp. They had lids, which, in use, were covered with parchment or cloth that was tied down around the neck to seal the contents. The jars had Hispano-Moorish designs—leaves, geometric patterns, vines, and flowers—in luster or they were tin glazed in blue and white or maiolica.

Albarello often had the name or, more likely, abbreviation of the contents painted on the front in Latin script. For instance, a jar now in the Getty Museum is labeled "syrupus acetositatis citriorum,"

Lovely Spanish Albarello (pharmacy jar), circa 1671–1730.

which translates as "syrup of lemon juice," believed to cure hangovers and vertigo as well as calm fevers. Apparently hangovers were the business of pharmacists of old.

By the early sixteenth century, apothecary wares had expanded to include vessels with spouts, bulbous jars for roots, and squat jars for syrups and oils. They were found throughout Europe.

Pharmacists also used pill slabs. These were flat tilelike squares or octagons of clay decorated in a manner similar to the jars, including, by this date, heralds and perhaps a Latin quote or slogan. Druggists, who made everything to order, used pill slabs to roll herbs and powders with gum arabic into pills. Pill slabs were also made with holes for hanging, so the druggist could hang it by the door to advertise his services.

Barber's bowls, also in widespread use at the time, were shallow, with a wide rim and a cutout for the customer's neck. They could be comfortably held under the chin during a shave. Barbers and men at home also kept ceramic mugs to warm shaving soap.

During the sixteenth century, adventurers and explorers returning from their travels in the New World brought home souvenirs of their trips, among them clay pipes and tobacco. Tobacco was thought to be healthful, even curative. It was prescribed for toothache, cancer, deafness, consumption, asthma, venereal diseases, kidney stones, gout, tonsillitis, nosebleeds, cataracts, wounds caused by arrows, and a host of other troubling ailments.[1]

Tobacco was cultivated and used by Native Americans long before the arrival of the Europeans. There is a Huron myth that in a time when the earth was not fruitful, the Great Spirit sent a woman to travel from place to place to restore health and abundance. As she walked, everywhere her right hand touched, potatoes sprung up, everywhere her left hand touched, corn grew. Finally, when corn and potatoes were plentiful and people's bellies were full, the woman sat down to rest from her labors. When she was refreshed, she stood, and where she had rested, tobacco appeared.

Native Americans smoked to relax, to seal friendship, and to pray. Wafting smoke from a pipe spoke to the Great Spirit. Pipes were smoked in ceremonies and in solitude.

Columbus was the first European to see and describe tobacco smoking. He mistakenly thought he was going to meet the Great Khan of China on what we now know as the island of Cuba when, instead of the Great Khan, he encountered indigenous Taino smoking tobacco firebrands. Subsequent explorers took note of the curious custom. The French, who observed the Micmacs smoking pipes in Nova Scotia, called the pipes "calumet," a name that has stuck.

Calumet pipes, or peace pipes, were used by Native Americans to smoke tobacco, usually mixed with bark or herbs, throughout North and South America. Pipes were made of reeds, wood, alabaster, stone, or clay. The favorite stone, pipestone, found in the upper Missouri River region of southwest Minnesota, was formed when the glaciers blanketed the northern tiers of the country. The weight of the ice, and the extreme heat of compression, turned the layer of red clay, sandwiched between two layers of quartz, into catlinite, or pipestone. Catlinite is reddish in color and takes a nice polish. It is clay naturally turned to rock; we can think of the glacier as Mother Nature's kiln.

The Spanish, French, Portuguese, and English all brought pipes and tobacco back across the Atlantic, where interest in the strange custom was piqued. At first only the wealthiest aristocrats could afford to purchase a pipe and tobacco to smoke. Pipe smoking became a status symbol. But by the seventeenth century, colonists were growing and exporting vast quantities of tobacco, and urban potteries were making pipes for smoking it. As the price of tobacco came down, the size of pipe bowls increased. Pipe smoking was now a popular fad. Soon, even common people could indulge.

By 1800 the manufacture of clay pipes was big business and it was very competitive. Many potteries made only pipes. In order to succeed, they had to differentiate their wares from other factories,[7] and they did this by embellishing their products in distinctive ways. Soon pipes sported botanicals, animals, portraits of famous people, guns, the names of taverns (where they might be given away free), listings for various societies and events, and bucolic maritime and countryside scenes. Bowls made in the shape of a human head were popular. Some pipes were outfitted with extremely long stems to cool the smoke.

As the Industrial Revolution surged through the factory potteries, pipes became so inexpensive that they were often discarded after a single use–they had become disposable items. They were so plentiful that a Victorian family might have several in each room of the house.

In addition to clay pipes, tobacco smokers required tobacco jars, or humidors, in which to store their tobacco. The potteries obliged and turned out lidded jars in myriad shapes. They were often glazed an appropriate glossy brown. Some potteries made tobacco jars in the shape of human and animal heads and, less often, full figures. Others made jars in the shape of buildings. Families kept their jars filled with tobacco on small tables within comfortable reach of the smoker's easy chair.

Potteries continued to produce clay pipes, but by the late Victorian period, wood and meerschaum became popular. Cigarettes and cigars were introduced. As the demand for ceramic pipes dwindled, potters began making very fancy glazed bowls and stems, which, instead of being cheap and disposable, they could sell for high prices in order to stay in business. After World War I, cigarettes took over the market and most of the remaining clay pipe manufacturers closed. By the 1950s, the health professionals realized that tobacco was not a curative, but instead caused heart and lung disease and death.

Today, smoking is out of favor in much of the world, but you can still purchase an inexpensive clay pipe, though not as readily as a century ago. Some studio potters are making art pipes and reproductions. Antique clay pipe collecting is a growing hobby.

Clay, particularly bentonite, is an important pharmaceutical ingredient. Bentonite is used as a filler and "due to its absorption/adsorption functions, it allows paste formation ... [and is consequently used in] industrial protective creams, calamine lotion, wet compresses, and antiirritants for eczema."[2] Bentonite is also used as "an antidote in heavy metal poisoning." Kaopectate, and similar products from pharmaceutical companies, prescribed for diarrhea, is made of kaolin and pectin.

Shops and websites catering to the herbal and alternative medicine market offer bentonite pills, capsules, and powders, some rich in calcium or iron. These promise to detoxify the intestines and promote regularity. Interestingly, eating clay—geophagia (earth eating)—which has been documented throughout the world, is considered a disorder. It has been disparaged since Aristotle coined the term, and many who indulge, do so secretly.

Women more than men, especially pregnant women, eat clay, but men too have been known to consume it. Clay is eaten in the countries of West Africa, in South America, in the American Southeast, and in Nepal and India. The clay is typically dug from a band well beneath the surface, or in some instances scooped from the wall of the house. It is dried and generally baked at a low temperature and "served" in chunks.

Clay eaters usually favor a particular kind of clay found in their region. In Georgia, women prefer white kaolin. Susan Allport writes that "four hundred to five hundred tons of eko, a clay from the village of Uzalla in Nigeria, were being produced each year and sold in the markets as far away as Liberia, Ghana, and Togo. Irregular blocks of these clays were sun-dried, then smoked and hardened for two to three days over a smoldering fire. In the process, they were transformed from their original gray shale color into the rich chocolate color and sheen of eko, the final product."[3]

In some countries, such as Nigeria, women keep chunks of clay in little pouches tied to their waists. Clay is openly sold in the marketplace. In the United States, clay eating is looked down upon and is generally done privately. In the Peruvian Andes, potatoes are dipped into clay slurry before eating.

We do not know why people eat clay. We do know that some animal species eat clay: parrots, gorillas, chimpanzees, elephants, bears, rats, and others. Scientist currently have two theories. One is that because clay contains minerals, it acts as a supplement, especially for pregnant or lactating women, whose requirements increase. The other theory is that the clay acts as a detoxifier. This too could be tied in to pregnancy because foods that might not be toxic to an adult might be toxic to a fetus.

In October 2003, the *Journal of Experimental Biology* accepted a paper describing the work of three scientists, Nathaniel J. Dominy from the United States, Estelle Davoust from France, and Mans Minekus from the Netherlands. In their research, the three used a model of the stomach and small intestine to test whether kaolin could act as a detoxifier in the stomach. They report, "Our results support hypotheses advocating an adsorptive function of ingested clay. For pregnant women the advantages of reduced toxicity and digestion-inhibition are clear. By adhering to gastrointestinal epithelia, clays may not only improve digestive efficiency, but also reduce fetal exposure to toxins tolerated by the mother. Similarly, economically disadvantaged children living in the tropics are also frequent consumers of soil. They are particularly susceptible to undernourishment and diarrhoeal dehydration, conditions that may be exacerbated by a reliance on marginal plant foods rich in tannins and toxins. It is notable that howling monkeys are geophagous when they consume foliage, which is often toxic."[4]

More research needs to be done. There are health problems associated with earth eating, such as intestinal blockage and even death. But most clay eaters consume only small, regular doses.

Clay is also used on the body to promote good health and beauty. Spas and herbal shops offer mud facials made with bentonite, kaolin, "Red Moroccan Clay," "French Green Clay," clay from the Dead Sea, and other specialty muds to cleanse, moisturize, soothe, and detoxify the pores, promising to leave you younger looking and more vibrant. It is believed that Cleopatra herself lathered her face with the mineral rich mud of the Dead Sea in order to preserve her famous good looks.

Clays mixed with herbs and spices, such as lavender and cloves, plus aloe or honey and other additives are sold in jars for those who wish to do their own facial. Spas mix their own secret mud concoctions and apply the mud to the guest's face while she relaxes. Generally, the clay, straight or with additives, is smeared in a thin layer over the face, starting at the center and leaving the eyes, nostrils, and mouth free. The clay is then allowed to dry for ten or fifteen minutes before being removed with a wet washcloth.

One does not need to limit oneself to a mud facial. Some spas also offer mud baths. Here, the guest is immersed from toe to neck in warm clay, and usually a facial is also applied at the same time. You can even enjoy this with a partner. Some Native American tribes may have indulged in mud baths too.

Do mud facials and mud baths do any good? Any clay worker can tell you that clay dries and exfoliates the skin and tightens the pores. Look at a potter's or mason's hands and you will usually see hands that are more wrinkled than the hands of a nonpotter or nonmason of the same age. But clay feels good to the touch. And a mud facial or mud bath would be an intermittent, luxury activity, not an everyday occurrence like making pots or laying bricks. If nothing else, mud is a wonderfully sensuous substance.

Environmentalists have turned to clay to detoxify the earth rather than their own bodies. Bentonite is used in wastewater purification systems. It is also used to seal landfills. The Koreans, Chinese, and Japanese have used clay sprays to fight red tides, algal blooms that contain harmful neurotoxins, since the 1980s. Red tide kills a large amount of sealife, including shellfish, fish, sea turtles, dolphins, and manatees, and it can trigger coughing fits in people. Of course, it is very detrimental to the seafood industry, with outbreaks costing millions of dollars, and tourism is hurt because no one wants to walk a beach littered with dead fish. When clay is sprayed into the infected ocean, the tiny clay particles clump together and trap the microscopic red tide cells. The cells then sink to the ocean floor, away from the various sea animals that they affect.

American environmentalists are not completely convinced that clay sprays are the answer; they worry that though red tide may be defeated, other environmental issues will arise from the amount of clay being added to the sea. Experiments are being conducted in New England, Texas, and Florida. Meanwhile, Korea and Japan claim success.

More mundanely, bentonite, with its adsorptive properties, is used to clarify beer, wine, and mineral water and to remove impurities from cooking oils such as canola oil. It is also used in detergents and liquid hand cleansers to remove solvents and as a fabric softener. And, perhaps most mundane of all, bentonite is used in kitty litter; because it clumps together, kitty's mess can be removed without emptying the whole litter box.

Bentonite has reportedly been used for an ominous—and potentially deadly—purpose as well. Tim Trevan, a former United Nations weapons inspector in Iraq, suggested in his 1999 book that bentonite was used in a one-step drying process to produce dried spores of anthrax.[5] It is unsettling to think that bentonite, a beneficial clay in so many ways, could be used for such a heinous purpose.

Clay also causes silicosis, or potter's lung. Dry unfired clay is dusty. Clay that is mined for use, whether commercially or by an individual, is dried, cleaned, and pulverized before mixing with water for use; it gives off dust. Bone dry pots are dusty. Spilled clay, slip, or throwing water that dries in the studio, workshop, or factory adds dust to the air. Powdered chemicals for glazes—feldspar, dolomite, ocher, etc.—are dust.

Without precautions such as damp mopping, air filters, respirators, and safety masks it is impossible not to breathe in the dust. For most of history, these precautions have been unavailable.

Dust, even of nontoxic materials, which most but not all pottery materials are considered, poses a health risk when inhaled. Silicosis, now officially classified as pneumoconiosis, occurs when fine dust in the form of free silica particles hangs in the air and enters the lungs. The lung capillaries then surround the particles, making new lung tissue. This causes the lungs to choke up and can cause death.

The coughing, shortness of breath, and more dire symptoms of potter's lung have been known for hundreds of years. Today, the disease is prevented by keeping the work area scrupulously clean and by wearing protection when mixing glazes or clay. In years past, it was considered "part of the territory" of working with clay.

Today's large-scale clay mines also pose environmental concerns. When potters dug their clay by hand, often from the same bed for generations, little damage was done to the earth. By the eighteenth century, English citizens were complaining about potter's clay pits and ordinances were passed to require that they be filled in. Now, heavy machinery is used to dig clay, often leaving gaping craters. In Georgia, where 60 percent of the world's kaolin is mined, disagreements and lawsuits have arisen between families who own the land and mining companies who claim the rights.

Nevertheless, though particular veins have been mined out, and problems exist, clay remains an abundant material. It adds far more good to our lives—and with common sense, our health—than it takes away.

ART, TOYS, GODS, GODDESSES, AND FERTILITY

Pottery is the most intense of the arts.
It brings the most to bear within the smallest compass.

—HENRY GLASSIE,
The Potter's Art

So Allah shaped Adam into a human being, but he remained a figure of clay for
40 years. The angels went past him. They were seized with fear by what they
saw, and Iblis felt fear most. He used to pass by the figure of Adam, buffeting it,
which would make a sound like pottery. Allah told us: "He created man (Adam)
from sounding clay like the clay of pottery."

—55:QURAN

G O TO A museum and wander through the halls and galleries.
Look at a storage vessel made by a Jomon potter four thousand
years ago, at the scrolled coils, the astonishing rim with its points and
dips, waves, and curlicues. Look at a red and black Greek figure
vase, the lean muscled man playing a flute, the soft gloss of the slip.
Look at a rotund salt-glazed wine jug from Germany, at a blue and
white Ming dynasty teacup, at a little female fertility figure, with
arms and fingers and breasts outlined with contrasting white clay.
Look at the little clay carts and horses and elephants from India.

Are these objects art?

They are displayed in glass cases, carefully illuminated and labeled. They are sought after by the most prestigious museums and wealthiest collectors. And each resonates with us. We respond to the beauty of the humble fertility figure, to the simplicity and clear colors of the teacup. Yet none of these objects was made as art; they were made for use in daily life, in ritual, for a meal, a celebration. To us, today in the twenty-first century, they are art. But to their makers they were just pots or figures.

For most of human history, everyone was a maker. A hunter chipped his weapons from stone. A woman made clay bowls so her family could eat. From earliest times, makers have valued skill and technical prowess. They have also yielded to the urge to embellish. There is no reason to scratch concentric bands around a jar or paint crosshatches with ocher on a beaker, except that a vessel decorated thus is pleasing to look at and the maker enjoys the act of decorating.

Even if a decorated object were not more pleasing to the eye than a plain one, and often it is not, often the beauty lies in the form, potters would still incise those lines and concentric rings around their vessels. Clay, more than any other medium, inspires manipulation, touch; it offers infinite possibilities for shape. But makers in other mediums too are beset with the desire to decorate, enhance.

Does that mean that an object that is embellished is art? Perhaps it is more intentionally art, but surely the overwrought and gilded soft-paste porcelain statuary of the eighteenth century is not more a work of art than the unglazed incense burners made by fifth-century potters working in Teotihuacán, Mexico.

The first works made of clay were figures, or possibly beads. Even after pots were made, figures continued to be produced. Indeed, many vessels became figures themselves. As we have discovered, the Venus of Dolni Vestonice, a small figurine, was made thirty thousand years ago, and her recovery pushed back the boundaries of what we know about clay making and perhaps religion. Small terracotta female figures appear throughout Eurasia with a stunning frequency, many, but not all, with exaggerated breasts and buttocks: a masked woman, her legs up in the birth position, from sixth-millennium B.C.E. Thessaly; a very pregnant female, holding the back of her head with one hand, her vulva with the other, about to give birth, from fourth-millennium B.C.E. Malta; flattish female figures with necklaces, widened hips, one with two spherical breasts, each with a pronounced vagina and birdlike face, from the Hurrian culture of the middle of the second millennium B.C.E. in Syria; a plump seated woman, her head lost long ago, hands in front of her abdomen, with full, perky breasts, from the fifth or sixth millennium B.C.E. in Iran. They have been found in Crete, the former Yugoslavia, and Hungary.

So-called Venus figures were made of stone, ivory, and bone as well as clay from Paleolithic times. The earliest figures were found largely in eastern Europe but occur throughout the continent and Russia. Anthropologist LeRoy McDermott explains, "These mostly palm-sized statuettes appear to depict obese women with faceless and usually down-turned heads, thin arms that commonly end or disappear under the breasts (but occasionally cross over them), an abnormally thin upper torso carrying voluminous and pendulous breasts, exaggeratedly large or elevated buttocks often splayed laterally but sometimes distended rearward, a prominent, presumably pregnant or adipose abdomen with a large elliptical navel, and what often appear to be oddly bent, unnaturally short tapering legs that terminate in either a rounded point or disproportionately small feet. Although readily recognizable, these anatomical details do not add up to an accurate image of the human figure."[1] He goes on to postulate that they represent women as they looked to themselves, not in a mirror, but looking down at their own bodies.

Today, we think of figures as sculpture, and sculpture as art. Most scholars do believe that these female figurines were goddesses or fertility talismans, symbolic of birth and regeneration. If it was women who made the earliest ceramics (which we believe but cannot know with absolute certainty), then it is not surprising that they would make small representations of an important aspect of their lives. Nor is it difficult to believe that the mysteries of birth, which brings a new little person from the womb, and the rejuvenation that comes with spring and the emergence of fresh green seedlings from the earth would become confluent in the Paleolithic and later Neolithic imaginations; infants and seedlings as renewal.

When things go awry, when there is no pregnancy or the baby is stillborn, when the spring rains do not fall and the soil turns to dust, when the seeds remain dormant in the earth and the grass refuses to green, one wishes for something different. It is understandable that hoping for a different outcome might lead to making a plea or offering to a higher power, a goddess, and that making a representation of her, before the event, might be a way of ensuring fertility, for both the woman and the earth.

Think of it; you are a young woman longing for your belly to swell like the bellies of the other women in your group. You watch as a baby emerges from between your sister's legs. You walk along the riverbank, alone. You sit in a cool, shady spot and, lost in your desires, you absentmindedly squeeze a ball of clay that you have scooped up from the shore. You pinch the ball here and there, not much, and before you know it, the clay has taken the shape of your dream, a miniature woman, bulging with a child in her abdomen, or your sister with her thighs up, ready to give birth.

Soon, the little clay figure not only embodies your wish; she becomes, in your mind, invested with power. In time, you teach your own daughter to make a goddess so that she will be fertile, so that the birth will not hurt too much, so that the child will be robust.

Some of the figures that archaeologists have recovered are mere suggestions of the female form, quickly made with a few pinches of the clay, the impression of a fingernail or stick, and the stroke of a finger. But many are more complex. Bits of clay were added to make breasts or arms or buttocks. Designs, age lines, even stretch marks were incised.

Archaeologists have also found terra-cotta birds, bulls (especially bulls' heads), pigs, and some male figures (though far fewer than females). In some locations, the animal figures predate the female figures. They have also found intriguing little models of one- and two-story buildings. Hundreds of these clay buildings, some better preserved than others, have been excavated, particularly in eastern Europe. They date from as early as the sixth millennium B.C.E. They have openings for windows or doors and, often, decorative lines. Some of the models sport the head or torso of an oversized woman on the roof.

Archaeologists disagree on the meaning of these miniature buildings. Many believe they represent dwellings; Marija Gimbutas and her followers believe they are likenesses of temples, and may have been used in rituals. Excavations have revealed that indeed, there were similar full-size buildings contemporaneous with the models. The little clay buildings do appear to be miniatures of the real thing—temples, workshops, houses, or a combination, sometimes complete with little ovens. Yet what the models were for remains a mystery.

Nevertheless, we can be reasonably certain that many of the woman figures were fertility objects. They may have been part of an earth-mother cult or goddess worship. As we have seen in Çatal Hüyük, the Anatolian city dating to 7000 B.C.E., ongoing excavations have uncovered terra-cotta figures with large, sagging breasts and swollen abdomens as well as figures only vaguely suggestive of the human face or form. They have also found bulls, birds, and some four-legged animals, which are difficult to confidently identify. Many of the houses appear to have interior shrines featuring bulls, which are also thought to be fertility symbols.

In the Abrahamic religions, instead of potters making goddesses of clay, the male God is the potter who makes a person of clay. The second chapter of Genesis states, "then the Lord God formed man of dust from the ground, and breathed into his nostrils the breath of life; and man became a living thing." In Hebrew, the word is "clay," not "dust." Indeed, "Adam" means red earth or clay.

Later, in Isaiah, God is literally likened to a potter: "Yet, O Lord, Thou art our Father; we are the clay, and Thou art our potter; we are all the work of Thy hand" (Isaiah 64:8 RSV) and later in Job, "The spirit of God has made me, and the breath of the Almighty gives me life . . . I too was formed from a piece of clay" (Job 33:4, 6 RSV).

In the Koran, Adam remains a clay figure for forty long years before he is given life. He even rings like a pot when passing angels bump into him.

In the Jewish legend of the golem, a servant is made of clay in the old German town of Worms, and he is given the task of saving the Jewish population from persecution by those who planted false evidence against the people and accused them of murderous rituals. After hearing a voice in the night tell him to make a man of clay, a rabbi and two pious men go to the riverbank on the outskirts of town to secretly do just that. Under cover of darkness, they fashion a full-size man of the sticky clay that is abundant along the shore. They stretch the statue out on the ground, like a man in repose, and then walk around him seven times. The golem glows bright red as if he were in a kiln fire. The devout men walk seven times in the other direction, and the golem begins to breathe. He does not speak, but he can see and hear and he does save the people from persecution.

In the Christian Bible, in the Book of John, Jesus and his disciples come upon a beggar who has been blind from birth. The disciples wonder who has sinned to make the man deserve such a horrid fate, and Jesus assures them that the man's blindness is not a punishment for anyone's sins, but a lesson. Then Jesus spits on the ground and makes "clay of the spittle." He covers the man's eyes with the clay and sends him off to a pool to wash. The man does as he is told. When he returns to the gathered crowd, to everyone's amazement, he can see.

Perhaps nowhere else in the world is clay more important to a people's spiritual well-being than in India. Clay vessels, lamps, and votive figures mark all of life's passages.

In northern India, the potter's wheel itself is an important part of the wedding rite. The ritual at the wheel ensures the bride's fertility and the success of the marriage. The women of the household go with the bride to visit the local potter. They bring powdered white rice, yellow turmeric, and red vermilion, which they use to draw sacred symbols on the wheel. The happy bride worships the wheel and then her relatives make food and monetary offerings to it (later the potter and his family enjoy these goodies). In some villages, the bride then sits on the wheel and is spun around seven times by the potter. In other villages, he spins the wheel while she watches. The potter's wife then gives the bride seven painted pots.

In Hindu weddings, pots are stacked, foot to mouth, to form towers, in the four corners of an enclosed space. The jars contain gifts of money, oil, grain, and such.

Terra-cotta elephants, horses, gods, and goddesses are important to the festivals and shrines of the tribals and Hindus of India. Both Muslims and Hindus commission potters to make earthenware elephants with clay bells and curlicues to celebrate the birth of a first son. The elephants are then kept in the house.

Most terra-cotta figures, however, are left outdoors at shrines set up along the edges of fields or under special trees. As years pass, they become covered with lichen, and finally, because of the low temperatures at which they are fired, they disintegrate back into the earth. At some shrines, there are generations of these votives— elephants, camels, musicians, horses, tigers, and bulls—in varying stages of decay, a quiet reminder of the fleeting nature of life.

Hindu villages each have their own shrines, and families have individual shrines dedicated to their ancestors. The clay votives are given in exchange for a request, as a promise, or as an appeasement. In addition to clay animals, sometimes clay replicas of particular parts of the anatomy are left at the shrine.

In Tamil Nadu in southern India, large terra-cotta horses, up to thirteen feet (four meters) in height are left at shrines for the spirit soldiers of the Hindu god Ayanaar to ride. Potters create the horses by building up thick coils of clay on the wheel and by beating, supporting the weight of the drying body with fired pots. Horses might be unadorned, abstract, stark, or they might be lifelike and lavishly and wildly embellished with clay decorations. Hundreds of these magnificent feats of the potter's art, some centuries old, stand together at the shrine at Urupetti. These votives are fired, but unfired figures are used in the spring festival of Rali Shankar.

To her shame and horror, Rali Shankar learns on the day of her wedding that she is marrying a younger man. She begs the gods to protect other young women from such a dreadful fate and then kills herself. For her festival, young girls purchase unfired clay figures of Lord Shiva and his wife, Parvati, from the local potter. They dress the figures in fancy clothes and jewelry and worship them. Then, after ten days of honoring the figures in this way, they marry them. After the wedding ceremony, complete with music and food, they take them to a nearby body of water, a stream or pond, and immerse them. Unfired, Lord Shiva and Parvati quickly dissolve. They came from the earth and they return to the earth. But next year, in the spring when it is time once again for the festival, Lord Shiva and Parvati will be "resurrected" by the potter and the cycle will begin again.

Similarly, in Bengal, for the February festival of Saraswati, the goddess of culture—literature, visual art, and music—potters make but do not fire a statue of the patron goddess. After two days of wrapping the mud Saraswati with flowers and offerings, her followers return her, like the figures elsewhere in the country, to the earth by plunging her into the river, where she dissolves. So too, in the state of Maharashtra, the elephant god Ganesha is honored with a clay likeness that, after the ceremonies, is given over to the river to dissolve.

These figurines are each made with deep care on the part of the potter, who calls upon years of practice and exacting skill. Nevertheless, none of these ritual objects, works of art, outlasts the short duration of the ceremony; they are consigned to the ravages of the churning river. The beautiful sculptures of the gods and goddesses become shapeless mud swirling in the murky water until it settles on the bottom.

In West Bengal, villagers commission large ceramic "trees" festooned with detachable cobra heads in honor of the snake goddess Manasa. Made from thrown and press-molded parts, the "trees" are often an imposing four feet in height (1.2 meters). They are fired in reduction, making them beautifully black.

In addition to tall clay trees and life-size ceramic horses, Indian potters are called upon to make miniatures. For the Hindu festival of Nava Khana, which celebrates the ripening of the rice crop, potters make hundreds of tiny pots, bowls for rice and water, little cooking pots with lids, miniature chula (stoves), grinding wheels, and little monkeys and oxen. After the festival, wheels are added to the monkeys and oxen. Now the miniatures become toys for the children.

Potters have probably always made toys for the children in their households. Clay invites play. Children play. And children underfoot need to be occupied and amused. We must wonder whether at least some of the humble figures that archaeologists have found, the little animals, perhaps some of the figures, might have been toys or dolls instead of gods and goddesses. It seems reasonable that a child of the tenth millennium B.C.E. would play with a little clay person as her mother nursed a newborn or prepared a meal. However, archaeologists disagree on whether dolls as toys or dolls as religious objects appeared first. Dolls, and perhaps spiritual relics, would have been made of other materials too—sticks, fur, leather, and feathers—but it is fired clay that endures.

In the nineteenth century, dolls with glazed porcelain heads and unglazed bisque heads (parian) were the cherished playthings of well-off little girls. Germany and France dominated the doll industry, with manufactories such as Dresden leading. The dolls came with extensive wardrobes of the latest fashions, plus sewing patterns so that children could make clothes themselves. Beginning around 1850, the French began making bébés, dolls that were children rather than adult women. Soon, baby dolls were more popular than the fashion dolls. At the turn of the century, dolls with bisque bodies—china dolls—became popular.

Early American potters made miniature jugs, churns, bean pots, dogs, and barnyard animals for their own children to play with, and later they made them to sell. They also made whistles and marbles as playthings.

Clay marbles have been made for thousands of years. Rolling a piece of clay between your palms to make a ball is one of the easiest things to do with soft clay and the first step in making a pinch pot. Ancient Romans and Egyptians used nuts and beach stones as marbles. They also made clay marbles, and in fact Romans considered playing with marbles a pleasant pastime.

In the eighteenth century, Americans made marbles of marble chips, but by the last quarter of the nineteenth century, they and Europeans were producing great quantities of inexpensive clay marbles. In the early twentieth century, however, machine production of glass marbles supplanted the manufacture of clay marbles, and today, nearly all marbles are made of glass.

Potters made rattles too, filling pots with tiny balls of clay or stones before closing them up and firing them. In our culture, rattles are most often toys for infants, but they are also used as rhythm instruments. In history too, rattles have had various purposes, sometimes spiritual or to bring the rain. In ancient Greece, rattles were often shaped like pigs' heads. The Yoruba make rattles shaped like people and shake them to ward off evil spirits.

Clay drums are less common, but in India an all-clay drum, shaped like a big pot, is played with the fingertips and palms. This drum, or ghatam, does not have an animal skin like most drums. Other clay drums do. Today, a few potters even make clay horns.

But it is clay whistles that are most widespread. They have been made throughout the world, for thousands of years, as playthings, for spiritual practices, for music.

The whistles that early U.S. potters made were whimsical toys, later produced as novelties for tourists. They were shaped like little jugs, gourds, and animals, and were made simply for fun.

Five thousand years ago, the hsuan, an egg-shaped ocarina or whistle, with up to eight finger holes, was played in Confucian rituals and in the royal court of China. Globular ceramic whistles appear in eastern and southern Europe, the British Isles, Russia, and Asia. In late-nineteenth-century Italy, Giuseppe Donati made earthenware ocarinas (Italian for "little goose") in various sizes. These egg-shaped flutes, with an air duct protruding from the base, became popular at carnivals. Later, in the 1930s, plastic versions, called "sweet potatoes," were all the rage in America.

Whistling pots were made for two thousand years in Latin America, from around 500 B.C.E. to 1500 C.E. when the Spanish conquistadors arrived. The Moche, the Chimú, the Inca, the Vicus, and others made them in Costa Rica, Colombia, Guatemala, Panama, and Peru. These whistling vessels usually had two chambers and a spout. Opposite the spout was the whistle. They were made in many imaginative shapes: monkeys, birds, people, jaguars, jars, and curious combinations, such as a monkey and a bottle.

Though the whistling vessels were common in the cultures of the Andes for two millennia we do not know their meaning or use.

This whistling vessel was made by a Chimú artisan sometime between 1000 and 1400 C.E. It produces an otherworldly sound.

Some scientists believe that they indeed held liquids. Others believe they were used to send signals or communications across vast distances. More recently, there is some thought that they were used in shamanistic rituals.

When several are blown at once, the sound is otherworldly, some would say mind altering. Dr. Isadore Rudnick of UCLA explained to author Daniel Statnekov that the reason for this phenomenon is "auditory beat," and added that "The sound that's perceived is comprised of both actual sounds and sounds that are self-generated due to the structure of our hearing mechanism. The effect is enhanced when the sounds are very loud. When a number of closely pitched whistles are superimposed in this manner, the effect is multiplied by each interaction."[2] It is possible that the Andean whistling pots were used to induce an altered state during religious ceremonies or as a healing ritual. Curiously, though there are illustrations of many activities on pre-Columbian ceramics, there aren't any illustrations of a whistling pot in use, or at least not that have been recognized.

The Moche (Mochica, 100–800 C.E.), who inhabited the northern coastal valleys of Peru before the Chimú and, later, Inca civilizations flourished in the region, made not only whistling vessels, but also portrait jars. Though many of the extant jars were found in tombs, they do not seem to have been made specifically for funerary purposes. However, like the whistling pots, we are uncertain of their meaning or use, though they do show signs of wear, such as abrasion and chipping.

The portrait jars often (but not always) have stirrup spouts. Stirrup jars were made in the area as early as 1500 B.C.E. and continued to be made, by succeeding peoples, until the Spanish conquest. What accounts for the three-thousand-year duration of the stirrup jar style is a mystery. They were seemingly prized, as potters spent considerable time and attention decorating them.

Christopher B. Donnan, the author of several books on the Moche and a professor at UCLA, which has an extensive archive of photographs of Moche art (much of which was looted before archaeologists could conduct proper excavations), writes, "True portraiture was among the greatest achievements of Moche potters. They skillfully captured the facial features of specific individuals and instilled a lifelike quality in each portrait.

"Nearly all of the Moche portrait head vessels depict adult males, although some children are also shown. No truly lifelike portrait of an adult female has been identified. Some portrait head vessels show individuals with illnesses, or with abnormalities such as a missing eye or a harelip.

"As a group, the portraits represent an astonishing range of physical types. They allow us to meet Moche people who lived more than fifteen hundred years ago and to sense the nuances of their individual personalities."[3]

Moche potters made the jars by pressing damp clay into clay molds. Once the pieces were stiff enough to remove from the mold and assembled, the potter added personalized touches, such a scar on a brow, a notable nose, and other identifying features. Some molds were used repeatedly with variations; some show the same man as he ages. The jars were burnished and painted with slips.

We know from the fine line drawings on Moche ceramics that they practiced a sacramental type of warfare. Rather than legions of

The Moche of ancient Peru made exquisite lifelike portrait jars. To look at one of these jars is to see the face of someone who lived between 100 and 800 C.E.

soldiers attacking garrisons or killing enemy combatants, the point of their almost sportlike events was to capture their opponents for humiliation and blood sacrifice. Professor Donnan and UCLA have in their collections a series of photos of portrait jars of a man they call Long Nose, because he has, well, a long nose. In one jar he is holding a bowl and spatula. In another, he has a cup. He is sometimes dressed as a warrior. But in one jar, he is naked, a rope around his neck, and we can surmise that he lost the battle and was captured himself, soon to be sacrificed.

The faces of the portrait jars are like the faces you see shopping in the grocery store today or walking down a city street. Except for the large earrings and some of the headdresses (though the head-scarves look suspiciously like the head coverings some fashionable male performers wear on stage today) they look like us.

In history, Moche portrait jars are without parallel. Nevertheless, potters, especially potters working on the wheel, which the Moche did not, are often seduced by the anthropomorphic qualities inherent in the jug or bottle form, and so, yielding to the sentiment, they give their pots faces.

Hiram Bingham found a pair of matching jars in Machu Picchu with two eyes and a toothy mouth painted on the neck, and a pointed nose stuck on. The jar has a swelling body and sturdy handle. Like the Moche, the Inca did not use the wheel. The face is abstract, almost like a child's drawing, but it is also captivating.

Sixteenth-century German potters, who earned much of their living making jugs for beer drinking—and likely enjoyed a brew themselves—expressed their great annoyance at Cardinal Bellarmine, the medieval priest who forbade drinking, by putting an unflattering image of him on the necks of their beer and wine jugs. These are called, yes, Bellarmines. They had actually been adding bearded faces to the necks of their brown, salt-glazed jugs for some time, but naming them Bellarmine jugs was sly revenge on the interfering father.

In the mid eighteenth century, Staffordshire potter Ralph Wood began making Toby jugs, kitschy representations of a sitting man in a tricorner hat, holding a mug of beer and often a pipe. Soon other Staffordshire potteries were producing Toby jugs. And then pot shops elsewhere in England and on the Continent began turning them out.

Vessels as faces or heads were made in the Roman empire featuring a muscled neck topped with a carefully modeled face, the chimneylike spout of the jug emerging from a head of wavy hair, so that the face itself was in the midpoint of the jug. The Santal tribe in India makes black-fired Bonga Devata, upside-down pots with grotesque faces, which they place under shrine trees. The nineteenth- and early-twentieth-century Mangbetu of the Congo made portrait bottles, cleverly using the neck of the bottle as the neck of the portrait, and placing the nicely sculpted head on top. Funerary effigy jars were made by Native Americans in the Mississippi Valley of Arkansas and by the Toltec in Mexico. The American South saw face jugs, which continue to be popular.

The jar as human head occurs throughout history and across geography. It has been executed with differing degrees of success. Potters have interpreted the form as realistic sculpture, as abstract representation, and as a fear-inducing work of ferocity. Face vessels have been used in religious ceremonies, as mortuary pots, and as a fun way to have a tipple. For many, however, we can only guess at the reason for their being or their use. We can, however, realize the widespread urge to make or own them.

Using clay to make representations of humans and animals—sculpture—has, as we have seen, occupied artists from the earliest times. In the academy, sculpture is generally considered fine art, though we must realize that the notion of art itself is rather recent. Fourth- and fifth-century B.C.E. Etruscans ornamented their temples with monumental terra-cotta sculptures of the gods. Tang dynasty Chinese made magnificent glazed sculptures of horses bristling with equine power and energy. Stylized terra-cotta sculptures of mothers and their children appeared in the Jenne culture of Mali (eleventh to seventeenth centuries C.E.) and the Akan/Fante culture of nineteenth-century Ghana.

Sculpture is universal. Interestingly, even sculptures made of other materials, such as gold, bronze, or silver, often start with clay. The "lost wax" method of forming molten metals, one of the oldest forms of metallurgy, dates from the third millennium B.C.E. It was used on virtually every continent except Australia, most notably in Africa and South America, and is still used by artists and jewelers today. The lost wax process utilizes clay and can be used to make hollow or solid sculptures.

For a solid object, a model is made of beeswax. This is covered with clay. In India, the first layer of clay might be fine kaolin, with successive layers of coarser clay added. Elsewhere, clay and dung might be used, or whatever clay is readily available. A pouring stem can be included in the wax model or added later by piercing the clay.

For a hollow object, a model is made in clay, which is then covered in layers of wax equal to the thickness of the desired metal wall. Again, a hollow pouring passage is created.

For both solid and hollow sculptures, the metal is heated to a molten stage and poured into the mold. The wax is "lost," or melted away, with the metal taking its place. When the metal is cooled, the clay mold is broken off and the metal is cleaned and polished.

Sculptors also sculpt in clay, keeping it damp for days and weeks as they work. This can be sent to a foundry, where a synthetic mold is made from the clay model and used for subsequent castings.

Metallurgy isn't the only art form or ancient technology in which clay has played a role. Neolithic weavers, and perhaps Paleolithic weavers, used fired and unfired clay weights in their craft.

Clay work is steeped in methods and traditions that span thousands of years. An artist working today would be as likely to make a small clay figure of a pregnant woman using exactly the same skills her Paleolithic predecessors used, even firing in the same way, as she would be to use a stainless steel needle tool to incise the lines, and a computer-controlled electric kiln to fire it. A man throwing on the wheel might make the same shape vase that his Song dynasty antecedent made. Yet because clay can take so many forms, it is so malleable and squishy, both metaphorically and in truth, it is a compelling vehicle for creativity. Despite the material's deep roots in the past, by its very nature it spawns artists who push the boundaries.

In the sixteenth century, French potter Bernard Palissy (1510–1590) stunned the public with his brightly glazed platters covered with high-relief snakes, frogs, lizards, fish, lobsters, shells, flowers, leaves, and vines. Nothing like these highly original trompe l'oeil dishes in combinations of bright greens, yellows, blues, and purples had been attempted or conceived of before. Palissy's plates were encrusted with amphibians and reptiles so realistic they looked as if they were alive, and his artifice of mixing animals, shells, and flowers in juxtapositions that would never be encountered in nature dazzled the public. Soon other ateliers were imitating Palissy's highly original works.

By the nineteenth century, however, factory potteries dominated production in much of the world. Individual potters who performed all the tasks themselves or in a small group—digging and pugging the clay (milling to a homogenous, workable consistency), throwing the pot, glazing and firing—struggled on in rural areas but lost most of their business to factories where workers performed specialized tasks. The conception of the work and the execution of it were now separate activities. In addition, other materials and different ways of life began to eliminate the need for much of the traditional potter's output.

Waves of artistic movements, the Arts and Crafts movement (1850–1910) in Europe and the United States and later the Art Nouveau movement, the Bauhaus in Germany (mid twentieth century), the Studio Potter movement (mid twentieth century to the present), and the Mingei movement in Japan (twentieth century) sprung up in reaction to the cold sterility of the Industrial Revolution. Each was an attempt to restore the integrity of the pieces being produced, of the maker and the made, and to forge a connection between the object, the material, and the artist.

William Morris (1834–1896), a socialist who was influenced by John Ruskin, founded the Arts and Crafts movement. Morris advocated beauty in the everyday furnishings of our lives: wallpaper, furniture, windows, fabric, and pottery. His message was taken up on both sides of the Atlantic. "Art potteries" sprung up to follow his message and many flourished.

In the United States, Cincinnati and the surrounding Ohio area was the heart of production of art pottery. It was here that a wealthy art school graduate, Maria Longworth-Storer (1849–1932), founded the famous Rookwood Pottery in the late nineteenth century. Longworth-Storer was initially interested in the then popular "feminine pastime" of china painting, but she soon grew beyond mere decorating, and learned to throw and to develop her own glazes. Rookwood operated for forty years.

In 1900, pottery entered the academy when the English potter Charles Fergus Binns (1857–1934) an advocate of the Arts and Crafts philosophy of William Morris, founded the ceramics program at New York State College of Ceramics in Alfred. Binns loved to experiment and test. He and his students worked in various clays and high and low temperatures.

One of his students was Adelaide Robineau (1865–1929) of Syracuse, who, like Longworth-Storer, began as a china painter. Robineau, best known for her intricately carved vases, worked in porcelain and developed stunning glazes, including notoriously difficult and often-elusive crystalline glazes. She taught classes at Alfred and in 1899 she began publishing the journal *Keramic Studio*, thus extending her influence on amateur potters emerging in the United States and elsewhere.

George Ohr (1857–1918), the "Mad Potter of Biloxi," began working around the same time, and his work is often considered a part of Art Nouveau, though it foreshadowed the wild and innovative later years of the Studio Potter movement. Ohr, who lived and worked in Mississippi, taught himself to throw and spent two years visiting other potters before setting up his own workshop, Biloxi Art Pottery. He developed brilliant multicolored glazes for his low-fired ware. What sets Ohr apart is his most unusual pots. He distorted the rims, added multiple handles and spouts, often turned them into private jokes or sexual innuendos. Ohr stashed most of his creations in a barn, unsold. Nine years before his death, he quit pottery altogether and became a Cadillac salesman!

In England Roger Fry (1866–1934), a member of the literary and artistic Bloomsbury Group, decided to learn to make pottery. He studied with a traditional country flowerpot maker and then, with fellow members of this freewheeling group, including Duncan Grant, Vanessa Bell, and Wyndham Lewis, set up the Omega Workshop to make, commission, or sell well-designed objects for the home. Their unfettered work, with its clean, simple lines, was a bit ahead of its time and the workshop never really succeeded commercially. It closed during the First World War.

It was another English potter, Bernard Leach (1887–1979), who ignited the modern Studio Potter movement. Leach, who was born in Hong Kong and studied art, was in Japan when he was exposed to the magic of pottery. Describing the momentous occasion, he wrote, "One day in 1911, two years after I had returned to the Far East, I was invited to a sort of garden-party at an artist friend's house in Tokyo. Twenty or thirty painters, actors, writers, etc., were gathered together on the floor of a large tea-room; brushes and saucers of colour were lying about, and presently a number of unglazed pots were brought in and we were invited to write or paint upon them. Almost all educated Japanese are sufficient masters of the brush to be able to write a decorative running script of, to Western eyes, great beauty, and many can paint. I was told that within an hour's time these pots would be glazed and afterwards fired in a little portable kiln, which a man was stoking with charcoal a few feet beyond the verandah in the garden. I struggled with the unfamiliar paints and the queer long brushes, and then my two pots were taken from me and dipped in a tub of creamy white lead glaze and set around the top of the kiln and warmed and dried for a few minutes before being carefully placed with long-handled tongs in the inner box or muffle. Although this chamber was already at a dull red heat the pots did not break. Fireclay covers were placed on top of the kiln, and the potter fanned the fuel til the sparks flew. In about half an hour the muffle gradually became bright red, and the glaze on our pots could be seen through the spy-hole melted and glossy. The covers were removed and the glowing pieces taken out one by one and placed on tiles, while the glow slowly faded and the true colours came out accompanied by curious sharp ticks and tings as the crackle began to form in the cooling, shrinking glaze. Another five minutes passed and we could gingerly handle our pots painted only one short hour before."[4]

Leach was smitten. He immediately set about finding a teacher. The type of pottery that he was making that day is called raku. It is highly prized for tea bowls for use in the Zen Buddhist–influenced tea ceremony practiced by tea masters in Japan. In the spiritual and ritualistic ceremony, each of the utensils—the vase for the flowers, the kettle for heating the water, the caddy for holding the tea, and most of all, the simple, handleless cup—are carefully chosen and treasured. The participants contemplate and revere the simple irregularities of the hand-formed pottery, the arch of a flower petal, the swirl of the tea, and the way the objects are handled is proscribed by the ceremony itself.

In the United States, raku has moved away from the tea ceremony; its appeal is more for the dramatic interplay of the fire, the pot, and the potter, the quickness of the process, and the lovely results that can be achieved in no other way. Americans have improvised and "improved" on the process, adding such postfiring treatments as plunging the red-hot pot into a bin of sawdust or newspaper or even dunking the glowing pot into cold water.

Leach found a teacher and became not only a potter, but an outspoken advocate for handmade pots and intensely opinionated about standards for excellence in pottery. With his friends the Japanese master potter Shoji Hamada and Michael Cardew (Leach's former student), he influenced several generations of potters in England, the United States, Australia, New Zealand, and Europe. Leach admired Japanese pottery and Song dynasty Chinese pottery. Cardew was fond of English country pottery. Leach's first book, *A Potter's Book*, has been in print since it came out in 1940. Cardew's, *Pioneer Pottery*, published in 1969, has been almost continuously in print.

In 1972, Bernard Leach adapted and published *The Unknown Craftsman*, written by his and Hamada's spiritual and intellectual friend, the religious philosopher Sōetsu Yanagi. Yanagi divided crafts into folk crafts and artist crafts, and subdivided folk crafts into guild crafts and industrial crafts, and artist crafts into aristocratic crafts and individual crafts. He most valued anonymous folk crafts. He coined the word *mingei*, which means "art of the people."

He wrote, "The special quality of beauty in crafts is that it is a beauty of intimacy. Since the articles are to be lived with every day, this quality of intimacy is a natural requirement. Such beauty established a world of grace and feeling. It is significant that in speaking of craft objects, people use terms such as *savour* and *style*. The beauty of such objects is not so much of the noble, the huge, or the lofty as a beauty of the warm and familiar. Here one may detect a striking difference between crafts and the arts. People hang their pictures high up on the walls, but they place their objects for everyday use close to them and take them in their hands."[5]

Few people working in clay in the Western world today embrace the idea or practice of working in anonymity. Most rebel at the idea that what they produce is not art. However, "the beauty of intimacy" is a self-evident truth for anyone in the industrialized world who is making everyday objects by hand or who uses them.

As the twentieth century deepened, the Studio Potter movement split between functional and sculptural. Peter Voulkos, a proficient thrower, tore the rims of his enormous platters, causing shock and excitement in the clay art world. Others followed his path, leaving behind the utility of well-formed casseroles or teapots for works that challenged widely held conceptions of what clay work should be.

Artists whose primary medium is not clay have been drawn to the versatility of the material. Pablo Picasso (1885–1973) became entranced with the possibilities of clay while on holiday in Vallauris in 1946. Visiting one atelier, he modeled two little bulls and the head of a faun. He returned the following year and in collaboration with the expert potter Georges Ramié, he made plates, jars, pitchers, and platters decorated with his trademark verve and spontaneity. Naively modeled fish swam on his platters, faces appeared on plates, vases, and bottles, and bulls and bullfighters reoccurred on tiles and plates.

Isamu Noguchi (1904–1988), best known for his large outdoor bronze and stone sculptures, worked in clay during three stays in Japan. He too worked with experienced potters. Noguchi made powerful, unglazed sculptures and also collected Japanese ceramics for inspiration.

Both functional and sculptural works are made in abundance in the twenty-first century, albeit with considerable crossover. Clay does not play the spiritual role in our lives that the Paleolithic goddesses played. We may nourish our souls when we serve our friends a meal on handmade plates. We might feel a spiritual uplift or catch our breath when a work particularly connects with us, be it ancient or new. But, unless we misunderstand the artifacts of history, clay objects are not as directly connected to our beliefs as they were to the Stone Age woman who modeled a Venus of mud.

Or are they?

There is an old church in Italy that is filled with specially made plaques. In the seventeenth century, a traveler stopped nearby for a drink of water. After refreshing himself, he absentmindedly left his cup behind. The local priest found the cup a few days later, and seeing that it was made of fine maiolica, with a lovely picture of the Madonna and child on the front, he placed it carefully in the crook of a venerable old tree near the church. Word of the tree with the

Madonna and child cup spread among the villagers, and one by one they came to see for themselves and to pray. They turned the gnarled old tree with the cup into an improvised shrine.

With so many worshippers coming to the tree, and with the evening breezes, it was inevitable that the cup should fall to the ground on occasion. But whoever was passing that way would pick it up and put it back into its resting spot in the tree. Then one day, when the cup fell, it struck a boulder and shattered. Shortly afterward, a man named Cristofano d. Francesco came to the little outdoor shrine to pray for his bedridden wife. He was sad and disappointed that the cup was no longer in its place. About to turn away, he noticed a fragment of the shattered cup stuck in a bush. It was the piece with the Madonna and child. Francesco took the shard and nailed it to the tree. Then he asked the Madonna to please cure his wife.

Miracle of miracles, the next day, she was smiling and well!

In gratitude, Francesco went to the local pottery and asked the potter to make a special plaque, which he gave to the nearby church in thanks. That was in 1657. Since then, many others have come to the place where the little cup was left and offered a plaque. Today, the little Church of Madonna dei Bagni has a rather large collection of plaques!

12

A FITTING DEATH
Urns, Gravestones, Companions, and Thieves

... dust to dust.

—BOOK OF COMMON PRAYER

FROM EARLIEST PREHISTORY, our ancestors have pondered the meaning of death.

Is there an afterlife?

If there is, what will we need for that life?

Does death begin when the last breath is exhaled or when the flesh has fallen away from the bones? How will we reunite our bodies with our souls in the afterlife? Will we need to? Does one need companions? Our ancestors fashioned answers for themselves, and they had to deal with the practicality of what to do with the body. Should it lie out for the vultures to peck clean? Does it require a house? How best to handle the decomposing flesh? Should the process be hastened? Should it be stopped, as with a mummy? Will it rise up again?

Most peoples find the notion that death might be permanent too awful to accept. Rituals, customs, and complex belief systems have evolved to reassure us that surely there is something else beyond death itself, something after life on earth ceases.

For millennia, in many different parts of the world, in different times, in diverse cultures, clay vessels played a role in these funerary rituals and the customs that grew up around the dead. Pottery appears frequently as grave goods at many ancient burial sites. In societies where cremation is practiced, urns are often used to hold the ashes of the dead. In some cultures, both cinerary urns and pottery for use by the deceased are interred together. Archaeologists are always pleased to uncover ceramic grave inclusions during the course of an excavation—they can discern a good deal of cultural information from the interred bowls or jars, including dates, external social influences on the society, and rank of the persons buried, though the purpose of the pottery is not necessarily obvious. In the graves of the most ancient cultures, it is difficult to discover with certainty whether the pots are for the use of the dead in an afterlife, or for the gods who must be pleased while the deceased makes a passage from one world to the next. Scholars generally agree, however, that pottery, stone tools, and other items interred with the deceased indicate a belief in an afterlife.

Writing of excavations in southern Greece, Sir Lindsay Scott pointed out that "it is significant to notice that in the great cemetery of rock-cut tombs at Mycenae, in which the citizens were buried, the vessels laid with the dead are entirely such as were used by the living. The only pottery objects peculiar to the tombs and belonging specifically to the ritual of interment were incense-burners and charcoal scoops."[1] In contrast, the contemporaneous Bronze Age shaft graves of the elite, also carved in rock and lined with stone, contained gold and silver, which were highly valued in Mycenae. In fact, heavy masks of gold covered the faces of the well-to-do dead. Still, even in these august sepulchers, there were stashes of ordinary clay vessels. Scott believed that these bowls and cups were for the meals that the deceased would need to prepare and eat in the other world.

Unfortunately, archaeologists studying the deep recesses of our past are not the only people interested in mortuary goods. When Heinrich Schliemann, famous for his discovery of Troy, excavated five of the six shaft graves located within the protective walls of the fabled Lion Gate at Mycenae, he was startled to find pottery of wildly varying dates jumbled up, the most ancient, from the early days of the city, mixed together with more recent examples. Upon further examination, he concluded that a thief had penetrated the graves during the end of the Mycenae era (the city was captured by the Argives in 468 B.C.E.). One of the three bodies had been disturbed, its clay and pebble covering removed, and it appeared that the bandit had grabbed only the more expensive objects, such as gold ornaments, the deceased's mask, and swords, and fled so quickly that he dropped smaller trinkets such as gold buttons in his haste.

We should not be surprised that a grave robber was active twenty-five hundred years ago. Surely as long as goods have been interred at burial sites, despite any taboos a culture might have, there have been people tempted to sneak in and take for themselves, or for the black market, what belonged to the no longer living. Fortunately, if the grave holds precious metals or jewels, the more common ceramics were often left behind, especially if the pieces were no different from what the burglar had in his own kitchen. Modern grave robbers, however, find a ready and lucrative market for ancient pottery, so much so that they are often called simply "pothunters" rather than named as the thieves that they are.

The stone monuments, cairns, dolmens, mounds, and caverns of Europe have attracted tourists, archaeologists, and modern-day pagans to their sites, each drawn by the secrets they hold. These are places closely associated with death. Many were built to shelter the bones of the deceased, and with the bones, artifacts, such as pottery.

Neolithic stone passage graves, long rock-lined tunnels buried beneath large earthen mounds, and variations carved into rock, or rocks heaped to form cairns can be found from the island of Crete, where the custom probably originated, to the shores of Ireland and Scotland. These tombs contained clay figurines as well as pottery bowls and vases. One interesting variation on these tombs is the "court" or "horned" tomb. These megalithic burial sites, found along coastal

Ireland, consist of a rock chamber with a patio-type court in front. Postholes encircle the tombs, and it is assumed that there was some sort of wood covering or fence. These date from the first half of the fourth millennium B.C.E. and, in addition to bodies, contain grave goods consisting of "decorated and undecorated pottery, arrowheads, scrapers, and knives."[2] The goods appear to be for the use of the interred.

In the bilevel Scaloria Cave in southeastern Italy, amateur archaeologists apparently removed much of the pottery, but fortunately, because the cave is so damp, stalagmites had formed, and some of the vases were stuck fast to the cave interior and could not be removed by even the most determined pothunter. Marija Gimbutas, who conducted an expedition there in 1979 and 1980, writes that consequently she collected shards from fifteen hundred vases from both the upper and lower caves, "decorated systematically with symbols of regeneration; eggs, triangles, snakes, plant and sun designs, and symbols of the Goddess of Regeneration herself–V's, triangles, hourglass shapes, and butterfly motifs."[3]

Gimbutas believed that in this case the cave, where 137 bodies were interred, piled in a heap, one upon another, was used for rituals as well as inhumation. The cave "consists of two separate parts; the top cave is a wide hall, suitable for habitation where ceramic and stone tools were found, dating from the end of the 7th to the end of the 6th millennium B.C.; the lower cave is long and narrow like a sleeve, with stalagmites and a live spring near the bottom. Near this spring many painted ceramic objects were found indicating, through radiocarbon dates, that ceremonies occurred there during the mid-6th millennium (c. 5600–5300 B.C.)."[4] Just as modern Judeo-Christian mourners might put flowers or a basket of holiday greens on the graves of loved ones at certain times of the year, or Roman Catholic mourners might say special masses for the deceased long after he or she has departed, these Neolithic peoples paid their respects by actually visiting the remains of the deceased and conducting rituals inside the grave with the bones, and on occasion, during that visitation, they'd bring a new bowl or ritual vessel as an offering. That modestly decorated clay bowl would be set down near the jumbled femurs, ribs, and skulls of the offeror's dead mother, sister, or child.

Gimbutas also discusses the Isbister Tomb of Orkney, Scotland, a megalithic structure covered over with a rock cairn and on top of that an earthen mound. Here it appears the interments include "disarticulated human skeletons, carcasses of sea eagles," the skulls separated from the body bones, plus a host of other bird skeletons. She notes, though, that the find is rare, despite the number of ossuaries in the chilly northern tip of Scotland, because in "the majority of megalithic graves, nothing has been found inside, since most were robbed in antiquity or destroyed by diggers."[5] However, the bird-cleaned skeletons, together with the remnants of ritual feasts and mortuary pottery, reminds us that burial or interment was far different for the Stone Age peoples of ancient Europe from what it is for ourselves today, with our sealed coffins, and often the word *death* not even spoken aloud.

Interment might take place immediately after death, but not necessarily. In many Neolithic and even Paleolithic cultures, and some later cultures, the body is left out for birds to remove its flesh before being buried. This was apparently the situation in the Isbister Tomb. In others, the body is interred, but removed later and reinterred elsewhere. The body might also be burned or partially burned in the tomb. But almost always, pottery, painted with slip drawings, was interred in the grave too.

On the other side of the globe, in North America, Native Americans practiced burial, cremation, and disarticulation. Often, they buried their dead with goods for the afterlife, or set bowls and clay smoking pipes and tools around the gravesite. The large mounds, found throughout eastern North America, but especially in Ohio and Wisconsin, were associated with funerals and with ritual. They were built from as early as 3000 B.C.E. to the sixteenth century of the Common Era. Though some appear to have been the foundation for large buildings or forts, others, such as the effigy mounds, shaped like birds and serpents, apparent only from the sky, appear to have contained burials including ceramics. There is some evidence that the pottery was deliberately and symbolically smashed at the time of burial rather than broken while interred.

Quite a few of these mounds were built over the course of

decades, growing in height as more corpses, as many as a thousand, were added. Some contained inner chambers, held up with wooden beams. Others contain cremations as well as the burials of unburned bodies. Archaeologists believe that the mounds played a deep spiritual role in the community as well as a mortuary role. They also required hard physical labor to build. Unfortunately, during the nineteenth century, adventurers, self-styled historians and pothunters opened many of the mounds and removed the pottery and metal goods, thus spoiling them for research. However, today a substantial number of remaining mounds are protected and scientific inquiry proceeds.

In Central America, Mexico, and South America, most of the inhabitants—the Inca, Maya, Mochica, Tarasco, and others—buried ceramics with their dead, although the pots themselves had different styles and the various cultures did not share the same religious beliefs or customs. The Spanish conquistadors ruthlessly destroyed so much of what they found when they arrived in the New World—books, pottery, homes, metalwork—that had the peoples of the Western Hemisphere not buried their pottery, most knowledge of these peoples would be lost to posterity. We might not even know that they were among the world's foremost clay workers.

In 1952, the Tomb of Pakal (603–83 C.E.) was discovered deep in the steamy jungles of the Yucatán region of Mexico by archaeologist Alberto Ruz Lhullier. The tomb was hidden inside the already well-known Mayan Temple of the Inscriptions. Pakal ruled the Mayan people from 615 C.E. until his death. During his reign, his city of Palenque flourished and grew. Eighty years old when he died, Pakal expected that he would have to descend into the underworld, but he believed that he would ultimately defeat death and be reborn. The magnificent Temple of the Inscriptions and his tomb were his preparation for his rebirth.

The temple sits high in the sky, up a flight of fifty-two stone steps. Inside is a secret door in the floor, hidden in plain sight for centuries. When Lhullier opened the door, he discovered a steep stone stairwell that descended eighty feet (twenty-five meters) to the carved stone-covered sarcophagus where the unusually tall Pakal still lay.[6] The stairwell was filled with pottery, jade, and rubble. The ceramic jars on the stairs contained the king's food offerings to the gods.

The Maya also made small wonderfully sculpted terra-cotta statues, called Jaina figures, after the island off the coast of the city of Campeche, where most of the figures were found. Some scholars think that at least some Maya traveled from elsewhere in the country to the island of Jaina specifically to make or acquire these extraordinary little sculptures and to bury their dead. Jaina figures include men with feathered headdresses that appear to the modern eye to be four sizes too large, and thin, wraparound loincloths; armed soldiers in top-heavy helmets; pretty women; old people; and young people. The Jainas were clay companions; they accompanied the deceased after death.

Although women do not usually appear in the marvelous illustrations or inscriptions on Mayan pottery, nor do they appear in stone, they are included among the Jaina figures, suggesting that women may have been the artists or mourners who made these lively ceramic sculptures. Whoever actually made the Jaina, they were proficient in ceramic work. The solid heads and hollow bodies were made in clay press molds. Capes, headdresses, belts, robes, facial characteristics, and other features were added to the molded body and the figure was then colored with slips. The figurines, which averaged about ten inches, were fired inside a broken pot to protect them from the vagaries of the flames, yet many acquired lovely fire clouds.

Another royal Mayan tomb, the Tomb of the Jaguar in Guatemala, was found exceptionally well stocked with large quantities of pottery and other goods. The pots, many in the straight-sided cylinder form favored by Mayan potters, include a beautiful black-and-red-striped vessel for an alcoholic cacao drink and pots covered with florid depictions of the Mayan gods. In other Mayan graves, the dead have been found interred with a plate placed over the face. The plates have a kill-hole, "a drilled or smashed hole in previously fired pottery that ostensibly released the powers that the object had accumulated through use"; however, "in this case they were probably designed to allow the departure of the soul or other vital essences rather than 'killing' the vessel itself."[7]

The Inca, who lived in what is now Peru, believed that the deceased would need a cache of supplies for their next life, and placed the necessities of daily living, including pottery for preparing and eating meals, into their graves. In Machu Picchu, the remote estate in the Andes that so captures the modern imagination, the royal household surrounded itself with retainers, many of whom appear to have come from far away and may have been artisans. Interestingly, they too are buried with a supply of household pottery, but the pots include both typical Inca designs and pots typical of their own homelands. Even in death, these workers retained some of their personal identity. They wanted the dishes they had grown up with to eat from in the afterlife.[8]

The Inca also practiced ritual sacrifice. A number of mummies of young girls, who were probably drugged or made very drunk before being buried alive, have been found. They too were interred with a wealth of treasures, including pottery. Much of the pottery in their graves is intriguingly doll-sized, miniature bowls and vases.

The Jomon, the talented potters who lived as hunter-gatherers in Japan for thousands of years and who made the earliest known clay vessels, buried pots with their dead. Their burial practices slowly evolved through their many centuries of island life. Adults were generally buried in pits, some lined with stones and possibly wood. A pit might be round, oval, or circular. Archaeologists believe that the occasional flask-shaped pits that have been uncovered were likely storage pits recycled into funerary use, further connecting, at least psychologically, food storage and storage of the remains of the dead.

In the Incipient and Late Initial Jomon periods, deep jars and stone tools, simple, everyday items, were placed in the pits with the deceased. In later periods, more lavish goods, including jewelry such as earrings, bracelets, beads, and waist pendants, were placed in the pit grave along with the traditional deep pottery jars and stone tools.

In the Late and Final Jomon periods, the graves were packed with goods. Now the deceased were supplied with clay figurines, stone and clay tablets incised with the outlines of the female form, shallow bowls, jars with narrow necks, earrings, bracelets, beads, waist pendants, deep jars, and stone tools.

Infants, fetuses, and young children up to six years old, but more often younger, were interred inside pottery jars, which were then buried near or in a house in an upright position. The jars either had a postfiring hole punched in the bottom or did not have a bottom at all, indicating they were specifically made for the purpose.

During the Late Jomon period, some adult bodies were also placed in jars. However, the jars were too small to hold an intact adult body, so interment had to wait until at least most of the flesh had decayed in a first-stage burial. The jars are found in proximity to stone-lined burial pits, so scientists believe the preliminary burials took place in the pits. Once the dead had been reduced to bones, they were removed from the pit and the bones were set into jars. It also appears that cremation was occasionally practiced.[9]

In the later Yayoi period (c. 250 B.C.E.–c. 250 C.E.) in Japan and in Korea, the dead were regularly interred in very large clay urns (sometimes stone), which were in turn buried and marked with a circle of stones or mound of earth, occasionally a dolmen. In the Grave Mound period (250–600 C.E.) that succeeded the Yayoi, the earthen mounds, or tumuli, erected over the graves of the rulers were massive and included a moat. During this period large clay funerary figures called haniwa were set in a ring around the mounds to act as guards and companions. The haniwa were coil built and, due to their size, hollow. Once they were in place around the funerary mound, all who approached could see them from a great distance. Haniwa included statues of men, women, fantastical or anthropomorphic animals, houses, boats, and horses. The human haniwa wore hats, earrings, necklaces, and a sort of knee-length tunic with long sleeves, over billowing pants, all finely sculpted in clay. Some of the figures were placed atop square or cylindrical pedestals or seats, making them even taller and more imposing.

In the Shang (c. 1751–1111 B.C.E.) and Zhou (Western, c. 1027–771 B.C.E.; Eastern, 771–221 B.C.E.) dynasties in China, people of importance were buried in specially constructed underground tombs furnished with bronze and pottery vessels filled with food for their meals in the afterlife. Sadly, unskilled amateur archaeologists and greedy robbers raided these graves and the graves of the royalty of later dynasties, many of which were discovered when railroads were first built across vast stretches of China. The magnificent mortuary pots that these unscrupulous grave robbers took from the tombs make up a large portion of the treasured Chinese ceramics now in museums or private collections. In the 1930s, as many as three thousand tombs were secretly pillaged during a period of heavy construction of roads and rail lines and the ceramic booty quickly disappeared into antique shops and collectors' hands.[10]

Cécile and Michel Beurdeley, experts on Chinese ceramics, write, in explaining ancient burial practices, "The tombs of the Western Zhou (Chou) were similar to those of the Shang with a sloping passage leading down to the mortuary chamber. Under the Eastern Zhou (Chou), the Chinese seem to have abandoned human sacrifice, or at least it became much rarer. The funerary pit, covered with a tumulus of rammed earth, consisted of a subterranean chamber; the wooden coffin was deposited directly over a hole where lay the body of a sacrificial animal, usually a dog. Some tombs had several chambers joined by corridors, in which chariots, arms, and the skeletons of animals have been discovered. The funerary furniture, the wine jars and vessels containing food were arranged in niches dug into the walls. In Henan (Honan), the Zhou (Chou) used thin bricks made in moulds for the construction of their funerary chambers. In Nanking, on the other hand, the tombs are simpler: the corpse is placed on a stone bed with the mortuary vessels round it, and the whole is covered with tumulus."[11]

In succeeding dynasties, exquisite ceramic figures were left in the mortuary chambers of the wealthy in addition to urns, bowls, and jars. The mountain tomb of a Han dynasty prince and his wife, which was a complete house for use in death, with multiple chambers and opulently tiled ceilings, contained hundreds of ceramic pots filled with delicious food and beverages plus many figurines.

These figurines were important to the dead, surrogate companions for the afterlife. The corpse of a five-year-old girl who died in 595 C.E. was found surrounded in her grave with "numerous figurines of soldiers and women in red slip covered earthenware."[12] A complete orchestra and two porcelain guards accompanied a deceased general, Zhang Cheng, in his tomb. In other graves, the fired clay companions include a female polo player, fierce men on horseback, a flute player, and a willowy figure with her hair pulled back, all compelling works of art.

The most impressive tomb sculpture is the famed terra-cotta army of the First Emperor of Qin, with over seven thousand larger-than-life-size fired clay statues of soldiers and two thousand fired clay statues of larger-than-life-size horses. Qin Shi Juangdi (260–210 B.C.E.) became king of the province of Qin when he was just thirteen. By 221 B.C.E. he had unified the thirteen provinces of southeastern China as the Qin dynasty and imposed standards for currency, writing, weights, and measures. He opposed Confucianism and strived ruthlessly to stamp it out, burning Confucian books and putting Confucian scholars and philosophers to death. He also ordered the destruction of all books pertaining to the past, keeping only treatises on medicine and science.

Work on Qin's tomb and terra-cotta army began almost as soon as he gained power and it continued for thirty-eight years after his death. The soldiers have solid legs and abdomens. The upper torsos were coil built and modeled and are hollow. The heads, also hollow, were press-molded and modeled. Each of the seven thousand soldiers has a different highly expressive face. Indeed, they represent the features of men from the various provinces. The feet, legs, and lower bodies of the horses were modeled from a solid chunk of clay. The rest of the body was coil built and hollow. The head, ears, and mane were attached while the clay was still soft. Vent holes were cut into the horses' bodies to prevent explosions in the kiln.

The feat of building and firing nine thousand larger-than-life-size clay statues is almost impossible for a modern ceramic artist to contemplate. Successfully accomplishing such a task two thousand years ago is, to use a cliché, truly mind-boggling. Each of the pieces is not only large, which requires great technical virtuosity both in building and in firing, but each piece is made with careful attention to the tiniest details. The underside of an archer's shoe has tread, the warriors' belts have perfectly represented buckles, and the clothes have folds. Nothing is left to the imagination; it is all there, sculpted in clay.

A project of such enormity was more than the imperial workshops could handle alone. Many of the warriors and horses were made at potteries elsewhere in the kingdom and signed with the first initial of the artist, before being shipped to the tomb site.

The army resides in three massive pits, which have been excavated and preserved. There is a fourth pit that was never finished and which did not receive horses or soldiers. Qin's immense mausoleum has been the subject of archaeological research for the past forty years, but the chamber where Qin's body rests has not yet been excavated. Ironically, he had the structure equipped with bows fashioned in such a way that if a grave robber penetrated, an arrow would automatically be released and kill him, yet the grave was penetrated within a few years of its completion.

None of the Egyptian pharaohs equipped his or her pyramid with life-size pottery men, but the tombs were stocked with treasure, much of it (especially in the Old and Middle Kingdoms) ceramic, including jars of food and clay models. Small figures, called ushabti, were placed inside the tomb to accompany the dead. These mummy-like figures were usually made of clay and glazed turquoise.

The tomb walls were covered with art depicting scenes from daily life, gardens, farms, and workshops. There are wonderful illustrations of Egyptian potters at work. One tomb painting from 300 B.C.E. shows the god Khum sitting at a potter's wheel. Instead of a vessel, a newly made person is emerging from the spinning clay. A tomb painting at Beni Hasan from about 1900 B.C.E. shows a Middle Kingdom potters' workshop, complete with kiln, a potter throwing at the wheel, an assistant wedging clay, and the execution of various other tasks common to a bustling pot shop. A wall painting in a tomb at Thebes dated circa 1459 B.C.E. shows a similar atelier, with one potter making adjustments to the kiln while another wedges a mass of clay with his feet and two others work at the wheel, one throwing, the other turning the wheel itself.

Bits of glazed pottery were sometimes used as an inlay in the coffin. Egyptians of means, however, often preferred precious metal and stone and left pottery to the less affluent classes.

Herodotus describes funerary customs in Egypt in his *Histories*. He says that when "a distinguished man dies all the women of the household plaster their heads and faces with mud, then, leaving the body indoors, perambulate the town with the dead man's female relatives, their dresses fastened with a girdle, and beat their bared breasts."[13] Covering one's face and head with mud as an expression of grief has to be one of the most direct associations of clay and death that history has seen.

■

CREMATION LIKELY BEGAN in Europe and the Near East around 3000 B.C.E. and spread across northern Europe. Dolmens, stone table-like tombs, appeared in the northwestern areas of Europe while megalithic ossuaries continued in the west. In both instances, disarticulated bones and burned or partially burned bones have been found.

In the late Bronze Age in England and Scandinavia, cremated remains were often placed in earthenware urns, a practice that spread from eastern Europe. The cinerary urns, which were identical to the well-made domestic urns then in use, were usually buried in humped barrows. They generally consisted of "barrel urns and bucket urns, decorated with finger-tip and finger nail impressions and horizontal cordons, and globular urns which seem to represent a fine ware element, made with finer fabrics, some with highly polished surfaces and more complex decorative elements."[14]

Puzzlingly, since the early Neolithic period, urns in England were made with a groove about three-quarters of the way up the vessel. This groove was carried down through the ages, becoming more pronounced until it became a heavy flange or cuff and finally an applied cordon. This evolving decoration, always in the same area of the vessel, dividing it into a one-quarter zone and a three-quarter zone, first appeared around 3500 B.C.E. and lasted until around 300 B.C.E.[15] It is unclear whether this design served a practical purpose or a symbolic one. Perhaps they were merely made this way because they had "always been" made this way, and no one questioned or dared deviate. These urns were often associated with funerals.

By the middle to late Bronze Age, cinerary urns were large

enough to contain the thighbones of a cremated man. Similar pots were apparently used for food preparation. Oddly, after the Bronze Age, the ability of the potters seems to have declined and pots grew smaller.

In the late Bronze Age, cinerary urns were placed in pit graves in flat cemeteries, or urn fields, though on occasion, cremation remains were committed to these cemeteries without the protection of an urn. Some of the larger urn fields were used for several hundred years; others held only a few urns, which were interred over a comparatively short period of time. Often, the urn field surrounded a barrow. Centuries of opening up the countryside with the farmer's plow, of digging and churning the earth for the erection of houses and commercial buildings and the expansion of roads and railways, have turned up many urn fields. However, because they lie beneath the earth rather than above, like mounds and cairns and megalithic tombs, it is impossible to know how many existed, what portion have been discovered, or how many have been destroyed.

It is tantalizing to think of the potters who made the urns, some for food to sustain life, some for the repose of the incinerated remains of their sisters and brothers. What did they think about? What did they believe? Nothing comes down to us but the pots they made, the finger marks on the clay, charcoal stains in the soil, bits of food and textile, and the cemeteries and mounds they left behind. Few of us get the opportunity to see one of the urns in situ, lift one in our hands, or peer inside at the charred remains, yet, thinking about the urns, we can sense our connection to the potter who shaped the clay and fired it, and the man or woman whose ashes are deposited within, people like us and not like us, who lived three thousand years ago.

In the mid nineteenth century, an Iron Age cemetery was discovered in Tuscany belonging to what came to be known as the Villanova culture after the nearby village. It is believed that the Villanova culture branched from the Urnfield culture in the tenth or ninth century B.C.E. The Villanova potters made cinerary urns shaped like helmets and simple two-piece conical forms. They also made little timber hut-urns, modeled after the wattle-and-daub pole houses in which they lived. These charming little hut-urns were usually round, but some square ones have been found. They were about ten inches high. They were made to house the deceased in the underworld.

Charming ten inch Iron Age cinerary urn modeled in clay and found in central Italy. This urn is located in Rome, in the Museo Prehistorico ed Etnografico.

The Etruscans, who united the villages of northern Italy in the seventh century B.C.E., and eventually superimposed their own acquired cosmopolitan ways, also practiced cremation. Their cinerary urns ranged from simple egg shapes to elaborate terra-cotta constructions, sometimes imitative of other regions and times. They believed that the soul stayed with the remains of the dead in their houses (urns) as long as they were happy. To ensure happy souls, they depicted pleasant occasions on the walls of their tomb chambers, where the urns were kept, and on the urns themselves.

The Etruscans also made face, or "canopic," urns, which featured a realistic likeness of the person whose ashes were held within. These splendid portraits influenced the Romans' artistic interests in later years. One exceptional double cinerary urn, called the Sarcophagus of the Spouses, dating from between 520 and 510 B.C.E., is a large polychrome terra-cotta box with lifelike full body sculptures of a man and wife reclining together and sharing wine and perfume. They appear to be having a rollicking good time together even though in reality, the loving couple is a pile of ashes inside.

Romans placed their cinerary urns in niches in the walls of large underground vaults or cellars called columbaria. The urns were made of clay and sometimes enclosed in a lead cylinder for extra protection from the dampness of the columbaria. Like other ancient peoples, Romans believed in an afterlife and took care with their funeral rites, which were held at night, to ensure a safe passage. On the night of the funeral, the family, friends, and professional musicians paraded to the outskirts of town, where they built a roaring funeral pyre. When the last flames had died down and the ashes had cooled, a close female relative would scoop them up and place them in the urn.

Cremation has been practiced in Hindu and tribal India for centuries but instead of using cinerary urns in which to preserve the ashes, Hindus scatter the ashes of the dead, letting them go back to the earth whence they came and to the place where their ancestors have returned. However, pottery plays a more significant role in the rituals of death than in most religions.

Jane Perryman, the British potter who has studied Indian potters and their work extensively and has written a particularly insightful book about the processes and beliefs of the potters there, *Traditional Pottery of India*, describes a funeral: "Ritual pots are an important part of the death ceremony and most potters keep a supply in stock. A small narrow-mouthed pot is filled with holy water and the eldest son of the deceased circumvents the funeral pyre carrying the pot on his shoulder, then smashes it on the ground before lighting the pyre. Once the body has been burned, a pot with a hole in its base is filled with milk and water and hung above the cremation site, the liquid slowly dripping out and purifying the ground. During the ten day period of mourning, it is believed in most Hindu families that the spirit of the dead person is in transition. On the tenth day, the eldest son breaks a pot to symbolically release the soul into its next life. Many households replenish all their earthenware vessels at a time of death in the family and in some areas several sets of cooking pots are used and replaced during the mourning period. At any time during the year severe personal problems can be attributed to the malevolent spirit of a dead relative which can be lured into a clay pot by a priest and trapped there by sacred mantras."[16]

Rather than commemorating the dead with something of permanence, Hindus purchase unfired figures from local potters and, in a ritual, let them disintegrate naturally and return to the earth, as the deceased have done.

In the Gujarat region of India, domed-shaped, lime-washed spirit houses are set out for the spirits of the dead. These unglazed earthenware pots, which are fired, have a square opening cut out of the front, and often a finial or second small dome on top.

By around 100 C.E. burial had replaced cremation in the Christianized world except under unusual circumstances, such as the plague. Jewish believers had always preferred sepulchers. With burial the prevalent practice, there was no longer a need for cinerary urns.

■

COMMEMORATIVE MONUMENTS, WHETHER placed directly on a grave, or as a marker in a town square, have been erected for thousands of years. There is evidence of sticks thrust into the ground in the urn fields of Bronze Age Europe. Piles of stones, dolmens, and upright stones have been used to mark the spot where a burial has occurred. In the Christian era, gravestones came to dominate.

In the American South, from the mid nineteenth century to the mid twentieth century, potters made cemetery vessels for their own families and for the people in their communities. Stoneware markers were also made as far north as Michigan and west to Texas. Though not as durable as granite, easily smashed by a graveyard vandal or nicked by a careless groundskeeper with a power mower, many have stood for a century or more. Usually salt-glazed, they took varied shapes, but the most common was an almost phallic version of an inverted large jug with the mouth closed, or series of stacked jugs, with the uppermost jug closed. Finials might be added, or the body might swell in and out in a series of ripples. Some potters got really fancy, especially for one of their own, and made markers with swags and appendages, and attached bottles. Occasionally, a marker took the shape of a typical stone marker, a slab with a disk on top. The names and dates of birth of the deceased could be written in cobalt or incised. Some markers were left plain.

Thrown grave markers, popular in the American South. Inscription: Nathan Cagle/Was Born March 25, 1856/Died December 1861/The Son of Landley/and Eeliza Cagle. Smaller markers are for the foot of the grave, and have no inscriptions.

American potters also made vases for the grave. These were cylindrical with a point at the base, which could be inserted into the soft soil of the gravesite. Glazed in a glossy brown slip, or salted, these simple forms, filled with water, could keep a bouquet fresh for several days.

In addition to grave vessels, a peculiar, arresting form of pottery appeared in the early-nineteenth-century South, particularly South Carolina, North Carolina, and northern Georgia. This was the face jug, made at first by slaves and copied later by white potters, and made now for the tourist and collector trade. The jugs had frightening faces, with broken bits of china embedded for teeth, and sometimes stones or crockery for eyes, exaggerated noses, and ears. Slaves, many of whom made bricks and pottery on the plantations where they were held, placed the jugs in burial grounds. Historians are uncertain of the purpose of the face jugs. One theory is that they were a carryover from an African custom; that they were placed on the gravesite for a year with the belief that if they broke during that time, it was a sign that the deceased was wrestling with the devil. Another theory is that they were left at the gravesite to ward off evil, and perhaps vandals too.

In 1874, British cemeteries were disease ridden and overcrowded. Because coffins had to be stacked one on top of another, a churchyard often reached the bottoms of the church windows. Sometimes old bones were removed under cover of darkness to make space for the newly dead. Queen Victoria's surgeon, Sir Henry Thompson, promoted the benefits of cremation with his book *Cremation: The Treatment of the Body After Death*. Anthony Trollope and others took up the cause, and ten years later, cremation was legal in England. A similar movement took place in the United States.

Today, cremations account for less than half of the funeral arrangements in the United States but there is a thriving business in urns. These vary from faux oriental porcelains to specially commissioned artist-made urns. You can even, where it is legal, have the ashes of your loved one turned into a glaze for a vase. Or the loved one might become an urn or a commemorative plate, with the ash mixed into the clay body in the way that cow bones were used to make bone china.

What would Sherlock Holmes think? Surveying the well-appointed living room of the deceased, perplexed that a body had never been recovered, yet convinced the poor dear was most definitely dead, he would be stumped. Watson might then point out to him that in fact the deceased was on the mantel.

The shimmering lavender glaze on the porcelain vase?

That, Sherlock, is the remains of our victim!

■

ARCHAEOLOGISTS RELY ON pottery more than any other artifacts to understand a culture, to place it in context with other cultures. Before carbon 14 dating, pottery was dated by which layer in an excavation it was found, and by comparing it with pots of known dates. Now the very reliable carbon 14 method, which is based on the half-life of carbon, is used, as well as archaeomagnetic analysis. Archaeomagnetic analysis utilizes the amazing fact that magnetic minerals, such as the iron in potter's clay or the mud of a hearth, "record the direction and strength of the earth's magnetic field at [the] location and times"[17] at which it was fired or burned. Of course, pots can be moved after they are fired in the kiln, so the location is only pertinent to hearths, but the strength of "frozen" magnetic markings is effective in dating pottery, as scientists have plotted the strength of the earth's magnetic poles over time.

Fired mud truly carries the record of civilization.

How a people handle their dead is a reflection of their beliefs about mortality and of their understanding of the health issues of a decaying corpse. Pottery, fired mud, has always been entwined with both the rituals and the practicalities of death; it has outlasted those it was meant to celebrate, to protect, or to guide.

Pottery does last an eternity.

Even when it is shattered, the shards remain.

For 'tis not verse, and 'tis not prose,
But earthenware alone
It is that ultimately shows
What men have thought and done.

—ALFRED DENNIS GODLEY

CONCLUSION

Who is the potter and who is the pot...

—EDWARD FITZGERALD,
Rubaiyat of Omar Khayyam of Naishapur

CLAY, FIRED AND unfired, has played an integral role in the march of civilization. It has given us shelter, from the simplest mud and stick huts of the early Stone Age to the multistoried apartments that "scrape" the skies of modern urban centers. It has been closely associated with food, with the farms and gardens where wheat and vegetables are raised, with the pots and stoves we use for cooking and the dishes from which we eat.

Indeed, perhaps clay's greatest role has been in the home. It has been made into fluted molds for puddings and cakes; colanders to drain rinsed lettuce from the garden; orange squeezers for the morning's juice; rolling pins for pie crust and pastry; cookie presses for Grandma's shortbread; fancy disks that you soak with water and leave in your bag of brown sugar to keep it soft; cheese domes; and all manner of mixing bowls and baking dishes. It has even been made into beds, and in China, into hollow, pillow-shaped headrests!

Clay enriches our holidays with ornaments for the tree; ice buckets for champagne; little villages of romantically old-fashioned houses that light up in the night; and candlesticks grand and small. Clay is the porcelain doorknob, the lamp pull, and the trivet. It has been turned into all manner of ingenious items, serving particular needs at particular times: the hot water bottle that our forebears placed at the bottom of the bed to warm the sheets in the days when sleeping rooms were cold; the two-part dish that, filled with water, keeps butter fresh without refrigeration. It is the binder in patchouli and sandalwood and other fragrant incense sticks. The buttons on our best coat might be made of clay, or the pink piggy bank in the bedroom, or the soap dish in the bathroom.

Clay has played an important role in our spiritual lives, enclosing the dead, celebrating the gods, and immortalizing our ancestors. We have seen that the lines between art, spirituality, and play are blurred. A clay whistle might be a toy, a sacred object, or an instrument for beautiful music.

And always, at least for thirty thousand years, clay figures have been with us. Fired and unfired. Realistic or stylized. Mud people have served as amulets, as gods and goddesses, as icons of belief. Mud people have stood guard over the dead and warded off loneliness in the afterlife. They have amused and entertained children as dolls, puppets, or parts of each. Sculpted, set out in museums or town plazas, or glazed and placed in a glass showcase, men and women and children, fashioned from clay, are our art.

The story of our relationship with clay is the story of material culture. It is the story of domesticity, and the story of technological advances. The inventions of the wheel and the kiln, the understanding that fire could turn mud to stone, were the foundation for the thousands of technologies that have followed.

Clay, as we have seen, covers much of the earth. Should our civilization crash, should a new dark age come, clay will still be here, and we can, each of us, scoop it from the earth, and use it for our most basic needs, the preparation of food, and for shelter.

Today, the scientists among us work to fashion new materials from the ingredients of clay. They experiment in labs and turn out knives that never dull and rocket ships to probe the solar system. Each day they have a new idea.

The story of mud is vast. Each of the chapters in this book could have been a book itself. Some of the topics would require an encyclopedia to cover in depth. My hope is that, reading this, you will lift your morning cup to your lips, and know that embodied in the cup are thousands of years of history. I hope that you think, sometimes, of all the anonymous people through the ages who have drunk from ceramic vessels just as you are drinking, of the unknown potters who fashioned images of mud and committed them to flames, of the Chinese who built walls and life-size armies and towering pagodas of the same material from which your cup is made.

And if you are a clay worker yourself, perhaps you will be inspired to make something new, a metaphoric urn field, or a teapot that changes color when filled. Maybe you will build an adobe tower with a clay bell to ring at sunset.

The greatest ceramic artist of all, of course, is Mother Nature. With the tiniest speck of clay, a mere particle floating in the air, she seeds the magic crystals we know as snowflakes. In her youth, she fired clay to make the ceramics we call rocks. She is generous to us with her beds of mud and has given us a great store of it, from which we can make our houses and fireplaces and dishes, and the white paper for our books, and the kitty litter for our cats.

Many believe that Mother Nature herself, or God, was a potter who fashioned the first humans from clay, that we ourselves are mud. We all know that, in death, our bodies "return to the earth."

HOW TO MAKE
YOUR OWN PINCH POT

The front of the thumb is an excellent tool.

—Paulus Berensohn,
Finding One's Way with Clay

PINCH POTS ARE one of the oldest and most versatile objects you can make from clay. You don't have to be an artist or a craftsperson to make one.

Anyone can do it.

You will need some clay. You can dig your clay yourself or you can purchase clay from a supplier. If you have clay near your home, you will need to test it to see if it is good for pot making. Dig up a piece about the size of a walnut. Clean out any sticks, roots, or stones. Roll the clay between your palms to make a snake. Now, curve this snake into a ring. If the ring doesn't crack, or cracks very little, you have good clay. If it cracks or crumbles, you will need to find a better spot to dig your clay. What you are looking for is clay that is smooth and plastic.

Now, once you have located your supply, dig a shovelful of clay.

Remove any sticks, roots, and stones that might be in your clay.

Most potters "wedge" their clay before beginning to make a pot. Wedging is like the reverse of kneading bread. The idea is to remove all the air bubbles from the clay. Wedging takes time to learn, but since you will be working with a small ball of clay that you are pinching, you can adequately wedge it by slapping it back and forth in your hands a half dozen times. Then you can begin.

Patting the clay, make a ball about the size of a plum.

Thrust your thumb into the center of the ball, but do not go through the other end. Using even pressure, and keeping your index finger and your second finger together on the outside of the wall, while your thumb remains inside, gently pinch the clay.

Turn the ball slightly, and pinch again.

Continue pinching until your fingers have gone all around the ball. Smooth the rim.

You now have a stout bowl.

Repeat this pinching and turning process, until the walls of your pot are just a bit over one-quarter inch thick or so.

Turn the bowl upside down and set it aside.

You can leave it like this, or you can make a foot.

To make a foot, roll out a coil of clay. Scratch a circle in the bottom of the pot that you have set aside. Mix a little bit of the clay that you dug, with water or, better, vinegar, to make a slurry. Put the slurry onto the scratched circle.

Now bend your coil into a ring the size of the circle that you scratched on the bowl.

Press the coil down onto the scratched ring, and smooth the edges.

Cover with a piece of cloth or plastic and let dry.

If you want to decorate your bowl, you can incise designs into it when it is "leather hard." That is, it has dried, but has not yet lightened in color. It will be the consistency of hard cheese.

Uncover your pot, and let it finish drying. You can tell if your pot is dry by holding it up to your cheeks. If it feels cool, it is not yet dry.

You can keep your pot in this unfired state if you wish. However, if you wish to fire your pot, you can use a simple outdoor fireplace made of a circle of stones, or you can dig a hole to make a pit.

Line either the pit or the hole with grass.

Set a handful of sticks down on the grass.

Place your pot in the center. Cover the pot with an armload of light, broken sticks.

Add more dry grass and more sticks. Set fire to the edges. You do not want the entire pile to burst into flame immediately. You want the fire to burn slowly and then take hold. The main reason pots break in bonfires like this is that they heat up too quickly. One way you can "cheat" is to preheat your pot in the kitchen oven. Start with a cold oven, and turn it up in intervals of 25° (less if your oven is Celsius).

When it has reached 500° F (260° C), grasp it with pot holders and carry it out to your waiting bed of grass and sticks. Cover it with sticks as above and proceed.

When your fire has cooled completely, you can remove your pot.

VARIATIONS

Beads

You can make beads by rolling balls of your clay between your palms or by making coils that you slice. Pierce your bead with a stick or drinking straw to make a hole for the string. Carve designs into the sides. You can make a round bead square by pinching between your fingers. Set aside to dry.

Fire in the same manner as your pinch pots. Beads are far less likely to break.

Rattle

Make two pinch pots. Make three or four beads the size of a raisin. Do not bother making a hole in the beads. Dust the beads with flour. Score the rims of your pinch pots by scratching. Mix up some slurry with additional clay and water or vinegar.

Put the slurry onto the rims of the bowls as if you were buttering them. Carefully, put the beads inside one of the pinch pots. Do not let any of the slurry get onto the beads.

Cupping the pinch pot with the beads in one hand, set the other pot on top, lining the rims up.

Press and smooth together. Clean up any slurry that squeezes out. Cover and set aside to dry.

When your rattle is leather hard, pierce it with a wooden skewer so that the hot air of the fire will be able to escape. Decorate if you wish. When completely dry, fire your rattle as your fired your pinch pot.

Enjoy!

MUSEUMS

Ceramic objects can be found in nearly every museum that holds collections devoted to history, art, archaeology, anthropology, material culture, natural history, science, or ethnography. Even most local historical societies have some ceramics on display. If you get into the habit of looking for the clay objects whenever you attend an exhibition or visit a museum or gallery, you will make many wonderful discoveries. Special exhibitions also often include clay works.

Clay objects, being three-dimensional, are best viewed in person. However, few if any museums allow you to touch their holdings and you must look at them through glass. Their websites, however, often feature multiple views of selected objects. There are also pieces shown online, including shards, which are not always on display. Many museums allow you to download or print images of their collections for your personal study, and some offer pdf versions.

Another wonderful way to see ceramics is to visit galleries, auctions, and antique shows. Here you can touch the pieces, so a visit, even if you are not in the market or the prices are more than you can afford, can be rewarding.

Following is a selection of museums with large collections.

Arizona State University Art Museum
Nelson Fine Arts Center
Tempe, AZ 85287
www.asuartmuseum.asu.edu

British Museum
Great Russell Street
London WCIB 3DG
England
www.thebritishmuseum.sc.uk

Brooklyn Museum
200 Eastern Parkway
Brooklyn, NY 11238
www.brooklynmuseum.org/

Canadian Museum of Civilization
100 Laurier Street, Station B
Gatineau, Quebec J8X 4H2
Canada
www.civilization.ca/archeo/ceramiq/cerart1e.html

The Everson Museum of Art
401 Harrison Street
Syracuse, New York 13202
www.everson.org

International Museum of Ceramics in Faenza
Museo Internazionale delle Ceramiche
Via Campidori 2
48018 Faenza
Italy
www.micfaenza.org

Fitzwilliam Museum
Trumpington Street
Cambridge CB2 IRB
England
www.fitzmuseum.cam.ac.uk

The Metropolitan Museum of Art
1000 Fifth Avenue (at 82 Street)
New York, NY 10028
www.metmuseum.org

Minneapolis Institute of the Arts
2400 Third Avenue South
Minneapolis, MN 55404
also
World Ceramics (virtual museum)
www.artsmia.org/world-ceramics/chooser.html

Museu del Càntir
Plaça de l'Església, 9 | 08310
Argentona (Barcelona)
Spain
www.museucantir.org

Peabody Museum of Archaeology and Ethnology
Harvard University
11 Divinity Avenue
Cambridge, MA 02138
www.peabody.harvard.edu/default.html

The Potteries Museum and Art Gallery
Bethseda Street, Hanley, Stoke-in-Trent
Staffordshire STI 3DW
England
www.2002.stoke.gov.uk/museums/pmag

Schein-Joseph International Museum of Ceramic Art
New York State College of Ceramics at Alfred
Alfred, NY 14802
www.ceramicsmuseum.alfred.edu

Smithsonian Institution
Washington, DC 20013
(includes sixteen museums, two in New York)
www.si.edu

Stoke-on-Trent-Potteries Museums
(Potteries Museum and Art Gallery, Gladstone Pottery Museum, Etruria Industiral
    Museum, and Ford Green Hall)
www.2002.stoke.gov.uk/museums/

University of Pennsylvania Museum of Archaeology
3620 South Street
Philadelphia, PA 19104
www.museum.upenn.edu/

Victoria and Albert Museum
Cromwell Road
London SW7 2RL
England
www.vam.ac.uk

NOTES

Chapter 1

1. Farmer, Fannie Merritt. *The Boston Cooking-School Cook Book*. Boston: Little, Brown, 1918; Barleby.com, 2000. www.bartleby.com/87/.
2. Toussaint-Samat, *History of Food*, 10.
3. Flandrin and Montanari, eds., *A Culinary History of Food*, 58.
4. Mitchell, trans., *Gilgamesh*, 185.
5. Tannahill, *Food in History*, 132.
6. Ibid., 63.
7. Cort, Exhibition Smithsonian's Freer Gallery of Art and Arthur M. Sackler Gallery.
8. Toussaint-Samat, *History of Food*, 206.
9. Weaver, *America Eats*, 160–63.
10. Halici, "A Kitchen in Sille," in Davidson, ed., *The Cook's Room*, 179.
11. Tilsley-Benham, "Morocco," in Davidson, ed., *The Cook's Room*, 173.

Chapter 2

1. Symons, *A History of Cooks and Cooking*, 24.
2. Singer and Holmyard, eds., *A History of Technology*, Vol. 1, 297.
3. Jacob, *Six Thousand Years of Bread*, 17.
4. Herodotus, *Histories*, Book II, 36.
5. Bresciani, "Food and Culture in Ancient Egypt" in Flandrin and Montanari, eds., *A Culinary History of Food*, 43.
6. Khalili, *Racing Alone*, 65.
7. Lyle, David, *The Book of Masonry Stoves*, 71.
8. Dick, *The Cottage Homes of England*, 109.
9. Ibid.

10. Ibid., 112.
11. *The Book of Masonry Stoves*, 103.
12. Ibid.
13. Wright, *Home Fires Burning*, 141.
14. Ibid., 136.
15. Cronon, *Changes in the Land*, 120–21.
16. Franklin, "An account of the New Invented Pennsylvania Fire-Place," quoted in Brewer, *From Fireplace to Cookstove*, 14.
17. Dick, *The Cottage Homes of England*, 113.
18. Rumford, "Of Chimney Fire-places, Essay IV (326)"
19. Tannahill, *Food in History*, 322.
20. Rumford, unpublished essay, Essay IV (404).
21. Symons, *A History of Cooks and Cooking*, 304.
22. Jacob, *Six Thousand Years of Bread*, 76–77.
23. Brewer, *From Fireplace to Cookstove*, 146–47.
24. Porcelain Enamel Institute, advertising brochure.

Chapter 3

1. Adams, *Korea's Pottery Heritage*, Vol. 1, 83.
2. Rhodes, Kilns, quoting Piccolpasso, *Three Books of the Potters Art*, 189.
3. Perryman, *Traditional Pottery of India*, 36–38.
4. Rhodes, *Kilns*, 13.
5. Perryman, *Tradional Pottery of India*, 59.
6. Shimada, "The Variability and Evolution of Prehispanic Kilns on the Peruvian Coast," in Rice, ed., *The Prehistory and History of Ceramic Kilns*, 110.
7. Adams, *Korea's Pottery Heritage*, Vol. 1, 29.
8. Medley, *The Chinese Potter*, 147.
9. Rhodes, *Kilns*, 47.
10. Childe, "Rotary Motion", in Singer and Holmyard, eds., *A History of Technology*, Vol. 1, 200.
11. Ibid., 202–3.
12. Medley, *The Chinese Potter*, 25.
13. Vincentelli, *Women Potters*, 14.
14. Childe, "Rotary Motion," in Singer and Holmyard, eds., *A History of Technology*, Vol. 1, 203.

Chapter 4

1. Freestone Ian, and David Gaimphes, *Pottery in the Making*, 104.
2. Cooper, *A History of World Pottery*, 40.
3. Medley, *The Chinese Potter*, 50.
4. Li and Cheng, *Chinese Pottery and Porcelain*, 6.
5. Ibid., 16.
6. Beurdeley, *A Connoisseur's Guide to Chinese Ceramics*, 23.
7. Medley, *The Chinese Potter*, 44.
8. Li and Cheng, *Chinese Pottery and Porcelain*, 16.
9. Beurdeley, *A Connoisseur's Guide to Chinese Ceramics*, 83.
10. Ibid., 83.
11. Li and Cheng, *Chinese Pottery and Porcelain*, 35.

12. Ibid.
13. Ibid.
14. Polo, *Travels of Marco Polo*, 250.
15. Hobson, *The Wares of the Ming Dynasty*, 168.
16. Ibid., 93.
17. Gleeson, *The Arcanum*, 68.
18. Gustafson, http://www.HotelMotel.com, "New Hotel Dinnerware Trends Are Designed to Please the Guest and Satisfy the Bottom Line," accessed April 19, 2004.

Chapter 5
1. McMurtie, *The Book*, 95–96.
2. Hobson, *The Wares of the Ming Dynasty*, 21.
3. Wilson, *The Scrolls from the Dead*, quoted in Hawkes, *The World of the Past*, 436.
4. Ceramic Tip Pen, Internet advertisement.
5. Hunt, *Historical Atlas of Ancient Mesopotamia*, 110.
6. King, *New York Times*, November 18, 2001.
7. Ibid.
8. Fiore, *Voices from the Clay*, quoting from the translation of L. W. King, 136.
9. Greenhut, "Ceramics for Paper" in Wachtman, ed. *Ceramic Innovations in the 20th Century*, 249.
10. Ibid.
11. Richerson, *The Magic of Ceramics*, 252

Chapter 6
1. Gimbutas, *The Civilization of the Goddess*, 64.
2. Schoenauer, *6,000 Years of Housing*, 90.
3. McHenry, *The Adobe Story*, 18.
4. Vitruvius, *The Ten Books of Architecture*, 1, 8, 20.
5. http://www.oldhousestore.co.uk, "Wattle and Daub by Ian Pritchett: An Explanation."
6. Bee, *The Cob Builder's Handbook*, 3.
7. Draper, *Dorset Country Pottery*, 23.
8. Ibid., 25.
9. Smith, *The Cobber's Companion*, 16.
10. Balter, *The Goddess and the Bull*, 33.
11. Lloyd, in Singer and Holmyard, eds., *A History of Technology*, Vol. 1, Chapter 17, "Building in Brick and Stone," 468.
12. Campbell, *Brick*, 33.
13. Ibid., 30.
14. Davis, *The Potter's Alternative*, 297–306.
15. Hobson, *The Wares of the Ming Dynasty*, 180.
16. Ibid.
17. Campbell, *Brick*, 70.
18. Ibid., 82.
19. Ibid.

20. Ibid., 92.
21. Wulf, *Traditional Crafts in Persia*, quoted in Porter, *Islamic Tiles*, 13.
22. Ibid.
23. Ibid., 63.
24. Tunick, *Terra-Cotta Skyline*, 25.

Chapter 7

1. Lambton, *Temples of Convenience and Chambers of Delight*, 11.
2. Horan, *The Porcelain God*, 48.
3. Lambton, *Temples of Convenience and Chambers of Delight*, 7.
4. Horan, *The Porcelain God*, 29 quotingFaber, Felix, History of Private Life.
5. Lambton, 18, quoting Mayhew, Henry, *Mayhew's London, Selections from "London Labour and London Poor,"* Ed. P. Quennell, London, *1949*.
6. Heine, Henrich as quoted http://www.plumbingproducts.com/hisplague.html10, Plagues and Epidemics from T&M Magazine; accessed February 3, 2002.

Chapter 8

1. http://www.fao.org.
2. http://sphakia.classics.ox.ac.uk.591/beeconf.nixon.html, accessed August 17, 2004.
3. Noël, *The Colonial Gardener*, quoted in Zug, *Turners and Burners*, 347.
4. Hobson, *The Wares of the Ming Dynasty*, 96.
5. Ibid., 58.
6. Greer, *American Stonewares*, 134.
7. Huxley, *An Illustrated History of Gardening*, 64.
8. Ibid., 65.
9. Ibid.
10. Evelyn, *Elysium Britannicum*.
11. Stonington Historical Society, "A Place to Take Root," exhibit catalog, Captain Nathan Palmer House, Stonington, CT, July/August 2004.

Chapter 9

1. *American Heritage College Dictionary*, fourth edition.
2. Singer and Holmyard, eds., *A History of Technology*, Vol. 1, Chapter 21, "Extracting, Smelting and Alloying," R. J. Forbes, 581.
3. Richerson, *The Magic of Ceramics*, 33.
4. Ibid., 34.
5. Ibid., quoting David W. Kingery, 2.
6. Lefteri, *Ceramics*, 42.

Chapter 10

1. Walter Reed Army Medical Center Online Patient Education, http://www.wramc.amedd.army.mil/education/tobaccocure.htm, accessed September 12, 2000.
2. International Minerals Association, North American, http://www.imana.org.
3. Allport, *The Studio Potter* 31, no. 2 (2003), "Women Who Eat Dirt," 79.

4. Dominy, Davoust, and Minekus, "Adaptive function of soil consumption: an in vitro modeling the human stomach and small intestine," in *Journal of Experimental Biology* 207: 319–24.

5. Trevan, *Saddam's Secrets*, quoted in the *New York Times*, "A Nation Challenged: The Bacteria: Officials Expanding Search, Warn Against Drawing Conclusions on Anthrax Search, Oct. 26, 2001.

Chapter 11

1. McDermott, "Self-Representation in Upper Paleolithic Female Figurines," http://cmsu2.cmsu.edu/ldm4683/2.htm, accessed August 22, 2004.

2. Statnekov, *Animated Earth*, 56.

3. Donnan, *Moche Portraits from Ancient Peru*, 9.

4. Leach, *A Potter's Book*, 30.

5. Yanagi, *The Unknown Craftsman*, 198.

Chapter 12

1. Scott, "Pottery," in Singer and Holmyard, eds., *A History of Technology*, Vol. I, 403.

2. Gimbutas, *The Civilization of the Goddess*, 210.

3. Ibid., 292.

4. Ibid.

5. Ibid., 293.

6. Miller and Martin, *Courtly Art of the Ancient Maya*, 207.

7. Ibid., 292.

8. Burger and Salazar, eds., *Machu Picchu*, 45.

9. Pottery in the Making, Pottery in Ancient Japan, 124.

10. Beurdeley, *A Connoisseur's Guide to Chinese Ceramics*, 39.

11. Ibid.

12. Ibid.

13. Herodotus, *The Histories*, Book II, 132–34.

14. http://www.eng-h.gov.uk/mpp/mcd/sub/bauf3.htm, accessed August 20, 2004.

15. Barton, *Pottery in England from 3500 BC to AD 1730.*

16. Perryman, *Traditional Pottery of England*, 69.

17. Barnett and Hoopes, eds., *The Emergence of Pottery*, 69.

BIBLIOGRAPHY

Ackerman, Andrew, and Susan Braunstein, *Israel in Antiquity* (New York: Jewish Museum, 1982).

Adams, Edward B., *Korea's Pottery Heritage, Vols. 1 and 2* (Seoul: Seoul International Publishing, 1990).

Allan, J. W., *Medieval Middle Eastern Pottery* (Oxford: Ashmolean Museum, 1971).

Allport, Susan, *Women Who Eat Clay*, Studio Potter 31, no.2 (2003) 79.

Andrews, Tim, *Raku: A Review of Contemporary Work* (Radnor, PA: A&C Black/ Chilton, 1994).

Atterbury, Paul, *Cornish Ware: Kitchen and Domestic Pottery by T. G. Green of Church Gresley* (Somerset, England: Richard Dennis, 1996).

Azoy, Mary Livingston, ed., *Peruvian Antiquities: A Manual for United States Customs* (Peru: OAS).

Baird, Daryl E., *The Extruder Book* (Westerville, OH: American Ceramic Society, 2000).

Baldwin, Cinda K., *Great and Noble Jar: Traditional Stoneware of South Carolina* (Athens: University of Georgia Press, 1993).

Balter, Michael, *The Goddess and the Bull: Catalhoyuk: An Archaeological Journey* (New York: Free Press, 2005).

Barnett, William K., and John W. Hoopes, eds., *The Emergence of Pottery: Technology and Innovation in Ancient Societies* (Washington, DC: Smithsonian Institution Press, 1995).

Barriskill, Janet, *Visiting the Mino Kilns* (Honolulu: Wild Peony/University of Hawaii, 1995).

Barton, K. J., *Pottery in England from 3500 BC–AD 1730* (South Brunswick, England: A. S. Barnes and Company, 1975).

Becker, Johanna, *Karatsu Ware: A Tradition of Diversity* (Tokyo/New York: Kodansha International, 1986).

Bee, Becky, *The Cob Builder's Handbook: You Can Handsculpt Your Own Home* (Murphy, OR: Groundworks, 1997).

Beirrel, Kenneth R., *Zen and the Art of Pottery* (New York/Tokyo: Weatherhill, 1989).

Berensohn, Paulus, *Finding One's Way with Clay: Pinched Pottery and the Color of Clay* (New York: Simon and Schuster, 1972).

Beurdeley, Michel, and Cécile Beurdeley, *A Connoisseur's Guide to Chinese Ceramics* (New York: Leon Amiel).

Bingham, Hiram, *Inca Land: The Incredible Story of the Discovery of Machu Picchu* (Washington, DC: National Geographic Society, 2003).

Birks, Tony, and Cornelia Wingfield Digby, *Bernard Leach, Hamada, and Their Circle* (Oxford: Phaidon/Christie's, 1990).

Bishop, Ronald, and Frederick Lange, eds., *The Ceramic Legacy of Anna O. Shepard* (Niwot, CO: University of Colorado Press, 1991).

Blair, Munroe, *Ceramic Water Closets* (Buckinghamshire, England: Shire Publications, 2000).

Boardman, John, *The History of Greek Vases* (New York: Thames and Hudson, 2001).

Borrelli, Frederica, and Maria C. Targia, *The Etruscans: Art, Architecture, and History* (Los Angeles: J. Paul Getty Museum, 2004).

Bottéro, Jean, *The Oldest Cuisine in the World: Cooking in Mesopotamia* (Chicago: University of Chicago Press, 2004).

Bourgeois, Jean, and Carollee Pelos, *Spectacular Vernacular: The Adobe Tradition* (New York: Aperture, 1989).

Branin, M. Lelyn, *The Early Makers of Handcrafted Earthenware and Stoneware in New Jersey* (London: Associated University Presses, 1988).

Brears, Peter C. D., *The English Country Pottery: Its History and Techniques* (Rutland, VT: Charles E. Tuttle, 1971).

Brewer, Priscilla J., *From Fireplace to Cookstove: Technology and the Domestic Ideal in America* (Syracuse: Syracuse University Press, 2000).

British Museum, *Timeline of the Ancient World: Mesopotamia, Egypt, Greece, Rome* (New York: Palgrave Macmillan, 2004).

Burger, Richard L., and Lucy C. Salazar, *Machu Picchu: Unveiling the Mystery of the Incas* (New Haven: Yale University Press, 2004).

Burney, Charles, and David Marshall Lang, *The Peoples of the Hills: Ancient Ararat and Caucasus* (London: Phoenix Press, 1971).

Burrison, John A., *Brothers in Clay: The Story of Georgia Folk Pottery* (Athens: University of Georgia Press, 1983).

Caiger-Smith, Alan, *Lustre Pottery* (New York: New Amsterdam Books, 1985).

——, *Pottery, People, and Time* (Somerset, England: Richard Dennis, 1995).

Camehl, Ada Walker, *The Blue-China Book* (New York: Halcyon House, 1916).

Campbell, James W. P., *Brick: A World History* (New York: Thames and Hudson, 2003).

Camusso, Lorenzo, and Sandra Bortone, eds., *Ceramics of the World: 4000 B.C. to the Present* (New York: Harry Abrams, 1992).

Carnegy, Daphne, *Tin-Glazed Earthenware: From Maiolica, Faience, and Delftware* (London/Radnor, PA: A&C Black/Chilton, 1993).

Caroselli, Susan, ed., *The Quest for Eternity* (San Francisco: Chronicle Books, 1987).

Carroll, Maureen, *Earthly Paradises: Ancient Gardens in History and Archaeology* (Los Angeles: J. Paul Getty Museum, 2003).

Casson, Lionel, *Libraries in the Ancient World* (New Haven/London: Yale University Press, 2001).

Clark, Garth, *American Ceramics: 1876 to Present* (New York: Abbeville Press, 1987).

——, *The Potter's Art* (New York: Phaidon, 1995).

Clark, Garth, Robert Ellison Jr., and Eugene Hecht, *The Mad Potter of Biloxi: The Art and Life of George E. Ohr* (New York: Abbeville Press, 1989).

Clark, Kenneth, *The Tile: Making, Designing, and Using* (Ramsbury, England: Crowood Press, 2002).

Cook, Scott, *Mexican Brick Culture in the Building of Texas 1800s–1980s* (College Station: Texas A&M, 1998).

Cooper, Emmanuel, *A History of World Pottery*, revised and updated (Radnor, PA: Chilton, 1988).

——, *A History of World Pottery*, 2nd revised edition (New York: Larousse, 1981).

——, *Ten Thousand Years of Pottery*, 4th edition (Philadelphia: University of Pennsylvania Press, 2000).

Cort, Allison, and Bert Winther-Tamaki, *Isamu Noguchi and Modern Japanese Ceramics: A Close Embrace* (Berkeley: University of California Press, 2003).

Coyle, Carolyn, *Designing with Tile* (New York: Van Nostrand Reinhold, 1995).

Cronon, William, *Changes in the Land: Indians, Colonists, and the Ecology of New England* (New York: Hill and Wang, 1983).

Cummings, Keith, *A History of Glassforming* (Philadelphia: University of Pennsylvania, 2002).

Daly, Greg, *Glazes and Glazing Techniques* (Kenthurst, Australia: Kangaroo Press, 1995).

Davidson, Alan, ed., *The Cook's Room* (New York: Harper Collins, 1991).

Davis, Harry, *The Potter's Alternative* (Radnor, PA: Chilton, 1987).

Degeorge, Gérard, and Yves Porter, *The Art of the Islamic Tile* (Paris: Flammarion, 2002).

de Lestrieux, Elisabeth, *The Art of Gardening in Pots* (Suffolk, England: Antique Collector's Club, 1990).

Denzer, Kiko, *Build Your Own Earth Oven: A Low-Cost, Wood-Fired Mud Oven* (White River Junction, VT: Chelsea Green Publishers, 2001).

de Waal, Edmund, *Bernard Leach (St. Ives Artists)* (London: Tate Gallery Publishing, 1998).

Dietz, Ulysses Grant, *Great Pots: Contemporary Pots from Function to Fantasy* (Madison, WI: Guild Publishing, 2003).

Dillingham, Rick, with Melinda Elliott, *Acoma and Laguna Pottery* (Santa Fe: School of American Research, 1992).

Dominy, Davoust, and Minekus, "Adaptive function of soil consuption: An in vitro modeling the human stomach and small intenstine," *The Journal of Experimental Biology* 207: 319-24, 2004.

Donnan, Christopher B., *Moche Portraits from Ancient Peru* (Austin: University of Texas Press, 2004).

Draper, Jo, with Penny Copland-Griffiths, *Dorset Country Pottery: The Kilns of the Verwood District* (Ramsbury, England: Crowood Press, 2002).

Eden, Victoria, and Michael Eden, *Slipware* (Philadelphia: A&C Black/University of Pennsylvania, 1999).

Emerson-Dell, Kathleen, *Bridging East and West: Japanese Ceramics from the Kozan Studio* (Seattle: University of Washington Press, 1995).

Eppler, Richard A., and Douglas R. Eppler, *Glazes and Glass Coatings* (Westerville, OH: American Ceramic Society, 2000).

Evans, Ianto, Michael Smith, and Linda Smiley, *The Hand-Sculpted House* (White River Junction, VT: Chelsea Green Publishers, 2002).

Failing, Patricia, *Howard Kotler: Face to Face* (Seattle: University of Washington Press, 1995).

Fanning, Janis, and Mike Jones, *Handcrafted Ceramic Tiles* (New York: Sterling, 1998).

Farmer, Fannie, *Boston Cooking School Cookbook*, revised (Boston: Little, Brown and Company, 1918).

Farrar, Linda, *Ancient Roman Gardens* (Phoenix Mill, England: Sutton Publishing, 1998).

Fehérvári, Géza, *Pottery of the Islamic World in the Tareq Rajab Museum* (Kuwait: Tareq Rajab Museum, 1998).

Fernández-Armesto, Felipe, *Near a Thousand Tables: A History of Food* (New York: Free Press, 2002).

Fiore, Silvestro, *Voices from the Clay: The Development of Assyro-Babylonian Literature* (Norman: University of Oklahoma Press, 1965).

Fischer, Steven Roger, *A History of Writing* (London: Reaktion Books, 2001).

Flandrin, Jean-Louis, and Massimo Montanari, *A Culinary History of Food* (New York: University of Columbia Press, 1999).

Flight, Graham, *Ceramics Manual: A Basic Guide to Pottery Making* (London: William Collins and Sons, 1990).

Fournier, Robert, *Illustrated Dictionary of Practical Pottery*, 4th edition (Madison, WI: Krause Publications/A&C Black, 2000).

Frankel, Cyril, *Modern Pots: Hans Coper, Lucie Rie, and Their Contemporaries* (Norwich, England: University of East Anglia, 2000).

Freestone, Ian, and David Gaimster, eds., *Pottery in the Making: World Ceramic Traditions* (London: British Museum Press, 1997).

French, Neal, *The Potter's Encyclopedia of Color, Form, and Decoration* (Iola, WI: Krause Publications, 2003).

Frith, Donald E., *Mold Making for Ceramics* (Iola, WI: Krause Publications, 1985).

Garfinkel, Yosef, *Dancing at the Dawn of Agriculture* (Austin: University of Texas Press, 2003).

Gately, Iain, *Tobacco: A Cultural History of How an Exotic Plant Seduced Civilization* (New York: Grove Atlantic, 2001).

Gimbutas, Marija, *The Civilization of the Goddess: The World of Old Europe* (New York: Harper Collins, 1991).

Giorgini, Frank, *Handmade Tiles* (Asheville, NC: Lark Books, 1994).

Girard-Geslan, Maud, *Of Earth and Fire: T. T. Tsui Collection of Chinese Art* (Seattle: University of Washington Press, 1999).

Glassie, Henry, *The Potter's Art* (Bloomington: Indiana University Press, 1999).

Glassner, Jean-Jacques, *The Invention of Cuneiform: Writing in Sumer* (Baltimore: Johns Hopkins University Press, 2000).

Gleeson, Janet, *The Arcanum: The Extraordinary True Story* (New York: Warner Books, 1998).

Goodwin, Jason, *A Time for Tea* (New York: Knopf, 1991).

Gorham, Hazel H., *Japanese Oriental Ceramics* (Rutland, VT: Charles E. Tuttle, 1971).

Graves, Alun, *Tiles and Tilework* (London, New York: V&A, Harry Abrams, 2002).

Greer, Georgeanna H., *American Stonewares: The Art and Craft of Utilitarian Potters* (Exton, PA: Schiffer Publishing, 1981).

Griffin, Leonard, *Clarice Cliff: The Art of the Bizarre* (London: Pavilion Books, 1999).

Guelberth, Cedar Rose, and Dan Chiras, *The Natural Plaster Book* (Grabiola Island, Canada: New Society Publishers, 2003).

Hamer, Frank, and Janet Hamer, *The Potter's Dictionary of Materials and Techniques*, 5th edition (London/New York: A&C Black/University of Pennsylvania, 2004).

Hawkes, Jacquetta, ed., *The World of the Past*, Vols. 1 and 2 (New York: Alfred A. Knopf, 1963).

Hayes, John W., *Handbook of Mediterranean Roman Pottery* (Norman: University of Oklahoma Press, 1997).

Heeney, Gwen, *Brickworks* (Philadelphia: A&C Black/University of Pennsylvania, 2003).

Held, Peter, ed., *A Ceramic Continuum: Fifty Years of the Archie Bray Influence* (Seattle: University of Washington Press, 2001).

Herodotus, *The Histories*, translated by Robin Waterfield (New York: Oxford University Press, 1998).

Hess, Catherine, ed., *The Arts of Fire: Islamic Influence on Glass and Ceramics of the Italian Renaissance* (Los Angeles: Getty Publications, 2004).

——, *Maiolica in the Making* (Los Angeles: Getty Research Institute, 1999).

Hildyard, Robin, *European Ceramics* (Philadelphia: University of Pennsylvania Press, 1999).

Hobhouse, Penelope, *Penelope Hobhouse's Gardening Through the Ages* (New York: Simon and Schuster, 1992).

Hobson, R. L., *The Wares of the Ming Dynasty* (Rutland, VT: Charles E. Tuttle, 1983).

Hopper, Robin, *Functional Pottery: Form and Aesthetic in Pots of Purpose* (Radnor, PA: Chilton, 1986).

Horan, Julie L., *The Porcelain God: A Social History of the Toilet* (London: Robson Books, 1996).

Howard, Jack, and Robin Hildyard, *Joseph Kishere and the Mortlake Potteries* (Woodbridge, England: Antique Collectors Club, 2004).

Hunt, Norman Bancroft, *Historical Atlas of Ancient Mesopotamia* (New York: Facts on File, 2004).

Huyler, Stephen, *Gifts of the Earth: Terracottas and Clay Sculptures of India* (Middletown, NJ: Grantha Corporation, 1996).

Jacob, H. E., *Six Thousand Years of Bread: Its Holy and Unholy History* (New York: Lyons and Burford, 1997).

Jones, Barbara, *Building with Straw Bales: A Practical Guide for the UK and Ireland* (Foxhole, England: Green Books, 2002).

Karmason, Marilyn G., with Joan B. Stacke, *Majolica: A Complete History and Illustrated Survey Updated and Enlarged* (New York: Harry Abrams, 2002).

Kaufmann, Gerhard, *North German Folk Pottery of the 17th to 19th Centuries* (Richmond, VA: International Exhibitions Foundation, 1979).

Kawami, Trudy, *Ancient Iranian Ceramics from the Arthur M. Sackler Collections* (New York: Harry Abrams, 1992).

Keeling, Jim, *The Terracotta Gardener* (North Pomfret, England: Trafalgar Square, 1990).

Kelly, Alison, *Mrs. Coade's Stone* (Upton-upon-Severn, England: Self Publishing Association, 1990).

Keswick, Maggie, and Alison Hardie, *The Chinese Garden: History, Art, and Architecture*, revised (Cambridge: Harvard University Press, 2003).

Ketchum, William, Jr., *American Country Pottery: Yellowware and Spongeware* (New York: Knopf, 1987).

Khalili, Nader, *Ceramic Houses: How to Build Your Own* (New York: Harper and Row, 1986).

——, *Racing Alone* (Hesperia, CA: Cal-Earth Press, 1990).

Kilroy, Roger, *The Compleat Loo: A Lavatory Miscellany* (London: Greenwich Editions, 1996).

King, Peter, *Architectural Ceramics for the Studio Potter* (Asheville, NC: Lark Books, 1999).

Kramer, Barbara, *Nampeyo and Her Pottery* (Albuquerque: University of New Mexico Press, 1996).

Kuroda, Ryōji, and Takeshi Murayama, *Classic Stoneware of Japan: Shino and Oribe* (Tokyo/New York: Kodansha International, 2002).

Lambton, Lucinda, *Temples of Convenience and Chambers of Delight* (London: Pavilion Books, 1997).

Lane, Peter, *Contemporary Porcelain: Materials, Techniques, and Expressions* (London/ Radnor, PA: A&C Black/Chilton, 1995).

Langendijk, Eugéne, *Dutch Art Nouveau and Art Deco Ceramics 1880–1940* (Rotterdam: Boijmans Van Beuningen Museum, 2001).

Latka, Tom, and Jean Latka, *Ceramic Extruding: Inspiration and Technique* (Iola, WI: Krause Publications, 2001).

Leach, Bernard, *Beyond East and West: Memoirs, Portraits, and Essays* (New York: Watson-Guptill Publications, 1978).

——, *Hamada Potter* (Tokyo/New York: Kodansha International, 1990).

——, *A Potter's Book* (Levittown, NY: Transatlantic Arts, 1970).

Lefteri, Chris, *Ceramics: Materials for Inspirational Design* (Mies: RotoVision, 2003).

Li Zhiyan and Cheng Wen, *Chinese Pottery and Porcelain* (Beijing: Foreign Languages Press, 1984).

Lungley, Martin, *Gardenware* (Ramsbury, England: Crowood Press, 1999).

Lyle, David, *The Book of Masonry Stoves: Rediscovering an Old Way of Warming* (Andover, MA: Brick House Publishing, 1984).

Manners, Errol, *Ceramics Source Book: A Visual Guide to the World's Great Ceramic Traditions* (Seacaucus, NJ: Chartwell Books, 1990).

Mansfield, Janet, *Contemporary Ceramic Art in Australia and New Zealand* (Roseville East, Australia: Craftsman House, 1995).

——, *Salt-Glaze Ceramics: An International Perspective* (Radnor, PA: A&C Black/ Chilton, 1991).

McConnell, Kevin, *Spongeware and Spatterware* (West Chester, PA: Schiffer Publishing, 1990).

McGovern, P., S. Flemming, and S. Katz, eds., *The Origins and Ancient History of Wine* (London: Routledge, 2004).

McGovern, Patrick E., *Ancient Wine: The Search for the Origins of Viniculture* (Princeton, NJ: Princeton University Press, 2003).

McHenry, Paul G., Jr., *The Adobe Story: A Global Treasure* (Albuquerque: University of New Mexico Press, 1996).

McKillop, Beth, *Korean Art and Design* (London: Victoria and Albert Museum, 1992).

McMurtie, Douglas C., *The Book: The Story of Printing and Bookmaking*, 3rd revised edition (London: Oxford University Press, 1943).

Medley, Margaret, *The Chinese Potter* (New York: Phaidon, 1989).

Miller, Mary, and Simon Martin, *Courtly Art of the Ancient Maya* (New York: Thames and Hudson, 2004).

Milner, George R., *The Moundbuilders: Ancient Peoples of Eastern North America* (London/New York: Thames and Hudson, 2004).

Minchelli, Elizabeth Helman, *Deruta: A Tradition of Italian Ceramics* (San Francisco: Chronicle Books, 1998).

Minogue, Coll, and Robert Sanderson, *Wood-Fired Ceramics: Contemporary Practices* (Philadelphia: A&C Black/University of Pennsylvania, 2000).

Mitchell, Stephen, trans., *Gilgamesh* (New York: Free Press, 2004).

Moeran, Brian, *Folk Art Potters of Japan: Beyond an Anthropology of Aesthetics* (Honolulu: University of Hawaii Press, 1997).

Morris, James, and Suzanne Preston Blier, *Butabu: Adobe Architecture of West Africa* (New York: Princeton Architectural Press, 2004).

Munsterberg, Hugo, and Marjorie Munsterberg, *World Ceramics from Prehistoric Times to Modern Times* (New York: Penguin Studio Books, 1998).

Nagatake, Takeshi, *Classic Japanese Porcelain: Imari and Kakiemon* (Tokyo/New York: Kodansha International, 2003).

Neils, Jennifer, ed., *The World of Ceramics: Masterpieces of the Cleveland Art Museum* (Bloomington: Indiana University Press, 1982).

Nelson, Glen C., *Ceramics: A Potter's Handbook*, 3rd edition (New York: Holt, Rhinehart, Winston, 1971).

Noble, Joseph Veach, *The Techniques of Painted Attic Pottery*, revised edition (New York: Thames and Hudson, 1988).

Obstler, Mimi, *Out of the Fire*, 2nd edition (Westerville, OH: American Ceramic Society, 2000).

Olsen, Frederick L., *The Kiln Book: Materials, Specifications, and Construction*, 3rd edition (Iola, WI: Krause Publications, 2001).

Osgood, Cornelius, *The Jug and Related Stoneware of Bennington* (Rutland, VT: Charles E. Tuttle, 1981).

Osterman, Matthias, *The Ceramic Surface* (Philadelphia: A&C Black/University of Pennsylvania, 2002).

——, *The New Maiolica: Contemporary Approaches to Colour and Technique* (Philadelphia: A&C Black/University of Pennsylvania, 1999).

Paul, Tessa, *Tiles for a Beautiful Home* (London: Merehurst Press, 1989).

Pegrum, Brenda, *Painted Ceramics: Colour and Imagery on Clay* (Ramsbury, England: Crowood Press, 1999).

Peirce, Josephine, *Fire on the Hearth: The Evolution and Romance of the Heating-Stove* (Springfield, MA: Pond-Ekberg Company, 1951).

Pereira, João Castel-Branco, *Portuguese Tile: From the National Museum of Azulejo, Lisbon* (London: Zwemmer Publishers, 1995).

Perryman, Jane, *Traditional Pottery of India* (London: A&C Black, 2000).

Peterson, Susan, *Contemporary Ceramics* (New York: Watson-Guptill, 2000).

Piccolpasso, Cavaliere Cipriano, *The Three Books of the Potter's Art* (London: Victoria and Albert Museum, 1934).

Polo, Marco, *Travels of Marco Polo* (New York: Modern Library, 1953).

Poole, Julia, *Italian Maiolica* (London: Fitzwilliam Museum Handbooks, 1997).

Porter, Venetia, *Islamic Tiles* (New York: Interlink Books, 1995).

Pounds, Norman, *Hearth and Home: A History of Material Culture* (Bloomington: Indiana University Press, 1993).

Quaknin, Marc-Alain, *Mysteries of the Alphabet* (New York: Abbeville Press, 1999).

Ramié, Georges, *Ceramics of Picasso* (Barcelona: Ediciones Polígrafa, 1985).

Reed, Cleota, *Henry Chapman Mercer and the Moravian Pottery and Tile Works* (Philadelphia: University of Pennsylvania Press, 1987).

Rhodes, Daniel, *Kilns: Design, Construction, and Operation* (Radnor, PA: Chilton, 1968).

——, *Pottery Form* (New York: Dover, 2004).

Rice, Prudence, ed., *The Prehistory and History of Ceramic Kilns* (Westerville, OH: American Ceramic Society, 1996).

Richerson, David, *The Magic of Ceramics* (Westerville, OH: American Ceramic Society, 2000).

Riddick, Sarah, *Pioneer Studio Pottery: The Milner-White Collection* (London: Lund Humphries, 1990).

Riley, Noël, *Tile Art: A History of Decorative Ceramic Tiles* (Seacaucus, NJ: Chartwell Books, 1992).

Robertson, Martin, *The Art of Vase-Painting in Classical Athens* (Cambridge: Cambridge University Press, 1992).

Robinson, Andrew, *The Story of Writing* (New York: Thames and Hudson, 1995).

Rodwell, Jenny, *Potter's Workshop* (Newton Abbott, England: David and Charles, 1999).

Rogers, Phil, *Ash Glazes* (London/Radnor, PA: A&C Black/Chilton, 1991).

——, *Salt Glazing* (Philadelphia: A&C Black/University of Pennsylvania, 2002).

Rorabaugh, W. J., *The Craft Apprentice: From Franklin to the Machine Age in America* (Oxford: Oxford University Press, 1986).

Sanders, Herbert, with Kenkichi Tomimoto, *The World of Japanese Ceramics* (Tokyo/ New York: Kodansha International, 1982).

Schoenauer, Norbert, *6,000 Years of Housing* (New York: W. W. Norton, 1981).

Schreiber, Tony, *Athenian Vase Construction: A Potter's Analysis* (Los Angeles: J. Paul Getty Museum, 1999).

Scullard, H. H., *Roman Britain: Outposts of the Empire* (New York: Thames and Hudson, 1979).

Seabrook, Charles, with Mary Louza, *Red Clay, Pink Cadillacs, and White Gold: The Kaolin Chalk Wars* (Atlanta: Longstreet Press, 1995).

Sentence, Bryan, *Ceramics: A World Guide to Traditional Techniques* (New York: Thames and Hudson, 2004).

Simpson, Penny, Lucy Kitto, and Kanji Sodeoka, *The Japanese Pottery Handbook* (Tokyo/New York: Kodansha International, 1979).

Singer, Charles, and E. J. Holmyard, eds., *A History of Technology, Vol. I: The Early Times to the Fall of Ancient Empires* (London: Oxford at the Clarendon Press, 1954).

Smith, Michael G., *The Cobber's Companion: How to Build Your Own Earthen Home* (Cottage Grove, OR: Cob Cottage, 1998).

Snell, Clarke, *The Good House Book: A Common-Sense Guide to Alternative Building* (New York: Lark/Sterling, 2004).

Speight, Charlotte F., and John Toki, *Hands in Clay*, 4th edition (Mountain View, CA: Mayfield Publishing, 1999).

Spivey, Richard L., *The Legacy of Maria Poveka Martinez* (Santa Fe: Museum of New Mexico Press, 2003).

Statnekov, Daniel K., *Animated Earth: A Story of Peruvian Whistles and Transformation* (Berkeley: North Atlantic Books, 2003).

Stewart, Dick, *The Cottage Homes of England* (New York: British Heritage Press, 1984).

Stuhr, Joanne, ed., *Talking Birds, Plumed Serpents, and Painted Women: Ceramics of Casa Grandes* (Tucson: University of Arizona Press, 2002).

Symons, Michael, *A History of Cooks and Cooking* (Urbana: University of Illinois Press, 2000).

Tannahill, Reay, *Food in History*, revised and updated (New York: Crown Publishers, 1988).

Teeple, John B., *Timelines of World History* (New York: Doring Kindersley, 2002).

Tichane, Robert, *Celadon Blues: Re-create Ancient Chinese Celadon Glazes* (Iola, WI: Krause Publications, 1998).

——, *Copper Red Glazes: A Guide to Producing These Elusive Glazes* (Iola, WI: Krause Publications, 1998).

Toussaint-Samat, Maguelonne, *History of Food* (Cambridge, MA: Blackwell, 1992).

Tregear, Mary, *Chinese Ceramics* (Oxford: Ashmolean Museum, 1987).

Troy, Jack, *Wood-Fired Stoneware and Porcelain* (Radnor, PA: Chilton, 1995).

Tung Wu, *Earth Transformed: Chinese Ceramics in the Museum of Fine Arts, Boston* (Boston: MFA Publications, 2001).

Tunick, Susan, *Terra-Cotta Skyline* (New York: Princeton Architectural Press, 1997).

Uglow, Jenny, *The Lunar Men: Five Friends Whose Curiosity Changed the World* (New York: Farrar, Straus and Giroux, 2002).

Van Lemmen, Hans, *Tiles: 1,000 Years of Architectural Decoration* (New York: Harry Abrams, 1993).

Van Lemmen, Hans, and John Malam, eds., *Fired Earth: 1,000 Years of Tiles in Europe* (Somerset, England: Richard Dennis Publications, 1991).

Van Lemmen, Hans, and Bart Verbrugge, *Art Nouveau Tiles* (New York: Rizzoli, 1999).

Vincentelli, Moira, *Women Potters: Transforming Traditions* (New Brunswick, NJ: Rutgers University Press, 2004).

Visser, Thomas Durant, *Field Guide to New England Barns and Farm Buildings* (Hanover, NH: University Press of New England, 1997).

Vituvius, *The Ten Books of Architecture* (New York: Dover, 1960).

Wachtman, John B., ed., *Ceramic Innovations in the 20th Century* (Westerville, OH: American Ceramic Society, 1999).

Walter, Josie, *Pots in the Kitchen* (Ramsbury, England: Crowood Press, 2002).

Warden, P. Gregory, ed., *Greek Vase Painting: Form, Figure, and Narrative* (Dallas: Southern Methodist University Press, 2004).

Watson, Oliver, *Ceramics from Islamic Lands* (New York: Thames and Hudson, 2004).

Weaver, William Woys, *America Eats: Forms of Edible Folk Art* (New York: Harper and Row, 1989).

Wescott, David, ed., *Primitive Technology II: Ancestral Skills* (Salt Lake City: Gibbs Smith, 2001).

Wheeler, Ron, *Winchcombe: The Cardew Finch Tradition* (Oxford: White Cockade Publishing, 1998).

Wilcox, Timothy, ed., *Shoji Hamada: Master Potter* (London: Lund Humphries, 1998).

Wilson, Richard L., *The Art of Ogata Kenzan: Persona and Production in Japanese Ceramics* (New York: Weatherhill, 1991).

——, *The Potter's Brush: The Kenzan Style in Japanese Ceramics* (New York: Freer Gallery of Art/Rizzoli, 2001).

Wing, Daniel, and Scott Alan, *The Bread Builders: Hearth Loaves and Masonry Ovens* (White River Junction, VT: Chelsea Green Publishers, 1999).

Wood, Donald, Teruhisa Tanaka, and Frank Chance, *Echizen: Eight Hundred Years of Japanese Stoneware* (Seattle: University of Washington, 1994).

Wood, Karen Ann, *Tableware in Clay: From Studio and Workshop* (Ramsbury, England: Crowood Press, 1999).

Wood, Nigel, *Chinese Glazes: Their Origins, Chemistry, and Recreation* (Philadelphia: A&C Black/University of Pennsylvania, 1999).

Wright, Lawrence, *Clean and Decent: The Fascinating History of the Bathroom and the Water Closet* (London: Penguin, 2000).

——, *Home Fires Burning: The History of Domestic Heating and Cooking* (London: Routledge and Kegan Paul, 1964).

Wulff, Han, *The Traditional Crafts of Persia* (Cambridge, England 1966).

Yanagi, Sōetsu, *The Unknown Craftsman: A Japanese Insight into Beauty*, revised edition (New York/Tokyo: Kodansha International, 1989).

Zakin, Richard, *Ceramics: Mastering the Craft*, 2nd edition (Iola, WI: Krause Publications, 2001).

Zeisel, Eva, *On Design* (Woodstock, VT: Overlook Press, 2004).

Zhang Wenli, *The Qin Terracotta Army: Treasures of Lintong* (London: Scala Books, 1996).

Zug, Charles III, *Turners and Burners: The Folk Potters of North Carolina* (Chapel Hill: University of North Carolina Press, 1986).

CREDITS

Pages 12, 16, 18, 65, 165, 167, and 169: Joseph Szalay.

Pages 21, 52, and 123: Jane Perryman.

Page 40: "The Bull, Cow and Calf" in *Graphic Illustrations of Animals Showing Their Utility to Man in Their Employment During Life and Uses After Death,* illustrated by Benjamin Waterhouse Hawkins. Printed by J. Graf. Published by Thomas Varty. Science and Society Picture Library, c. 1845.

Page 48: Alexander Marshak, courtesy of Eliane Marshack.

Page 59: From P. E. Newberry, "Beni Hasan," Part I, P1, 32, London, Egypt Exploration Society, 1893, from *A History of Technology.*

Page 60: T. A. Graves, from *A History of Technology.*

Page 63: Kunstkammer, Kunsthistorisches, Museum, Vienna.

Page 87: V & A Museum.

Page 90: Postcard (public domain).

Page 106: Phoebe A. Hearst, Museum of Anthropology.

Page 122: Courtesy of Matt McClure.

Page 124: Drawing by D. E. Woodall, based on P1 X. 1, E. J. H. Mackay, "Early Indus Civilization," 2nd edition, 1948, London, Luzac from *A History of Technology.*

Page 128: Eileen Griffin.

Page 135: Colored etching by Henry Alken, published by Thomas McLean of Haymarket, London Science and Society Picture Library, c. 1824.

Page 136: Science and Society Picture Library.

Page 147: Fotomax Index.

Page: 157: Courtesy of Monroe Blair

Page 184: Dong Lin, courtesy of California Academy of Science.

271

Page 194: Picture Library.
Page 213: Ed Chapell photo.
Page 215: Christopher B. Donnan, University of Texas Press.
Page 239: Drawing by E. E. Woodall for *A History of Technology*, Oxford.
Page 242: Collection of the Mint Museum.